Education in the Asia-Pacific Region: Issues, Concerns and Prospects

Volume 74

The purpose of this Series is to meet the needs of those interested in an in-depth analysis of current developments in education and schooling in the vast and diverse Asia-Pacific Region. The Series will be invaluable for educational researchers, policy makers and practitioners, who want to better understand the major issues, concerns and prospects regarding educational developments in the Asia-Pacific region.

The Series complements the Handbook of Educational Research in the Asia-Pacific Region, with the elaboration of specific topics, themes and case studies in greater breadth and depth than is possible in the Handbook.

Topics to be covered in the Series include: secondary education reform; reorientation of primary education to achieve education for all; re-engineering education for change; the arts in education; evaluation and assessment; the moral curriculum and values education; technical and vocational education for the world of work; teachers and teaching in society; organisation and management of education; education in rural and remote areas; and, education of the disadvantaged.

Although specifically focusing on major educational innovations for development in the Asia-Pacific region, the Series is directed at an international audience.

The Series Education in the Asia-Pacific Region: Issues, Concerns and Prospects, and the Handbook of Educational Research in the Asia-Pacific Region, are both publications of the Asia-Pacific Educational Research Association.

Those interested in obtaining more information about the Monograph Series, or who wish to explore the possibility of contributing a manuscript, should (in the first instance) contact the publishers.

Please contact Lay Peng, Ang (e-mail: laypeng.ang@springer.com) for submitting book proposals for this series.

Rupert Maclean, University of Tasmania, Hobart, Australia

Philip Wing Keung Chan, Monash University, Australia

Peter Edward Doe

Prospects for Professional Engineering Education in the Asia-Pacific Region

A Historical Case Study of the University of Tasmania

Springer

Peter Edward Doe
School of Engineering
University of Tasmania
Sandy Bay, TAS, Australia

ISSN 1573-5397 ISSN 2214-9791 (electronic)
Education in the Asia-Pacific Region: Issues, Concerns and Prospects
ISBN 978-981-96-3196-4 ISBN 978-981-96-3197-1 (eBook)
https://doi.org/10.1007/978-981-96-3197-1

This Springer imprint is published by the registered company Springer Nature Singapore Pte Ltd.
The registered company address is: 152 Beach Road, #21-01/04 Gateway East, Singapore 189721, Singapore

Foreword

I first encountered Peter Doe in 1965, following my enrolment as an undergraduate professional engineering student at the University of Tasmania (UTAS), when Peter had just commenced as a UTAS lecturer in mechanical engineering.

Following my graduation in 1968, I commenced a 45-year career with engineering consultancy Gutteridge Haskins and Davey (GHD) whilst Peter continued his career development in the UTAS School of Engineering, completing a PhD and was promoted to Associate Professor of Mechanical Engineering in 1991. Peter contributed widely to the advancement of the education of engineers not only in Australia but also in Southeast Asia and the Gulf States.

This book is a must-read for students of engineering education in that it is a case study of how to impart knowledge and skills (and what to avoid) based on his own experiences at UTAS. Peter traces Tasmanian engineering education as far back as the Zeehan School for Mines in 1894; then through to the establishment of the Faculty of Engineering in 1922 on the Domain, and to the relocation to the current Sandy Bay campus in 1957. Peter also discusses the proposal for a possible move to the Hobart CBD, which has been recently under consideration, and his preference for the ongoing retention of all the Science, Technology, Engineering and Mathematics (STEM) courses and facilities at Sandy Bay.

The book also focuses on the benefits arising from the encouragement of students from countries to Australia's north participating in Australian engineering education—from the days of the Colombo Plan through to the joint campus initiatives initiated by Australian universities more recently.

The UTAS Professor of Civil and Mechanical Engineering, Arch Oliver (also one of my lecturers) noted 'Engineers see the world differently from scientists and their training should reflect this difference. Whereas the scientist seeks to understand the reason for what is observed, the engineer is more concerned with providing an outcome in the face of nature's challenge'. Engineers are the unsung heroes of the professional classes in Australia, with the general population taking for granted the essential infrastructure and health benefits such as clean water, effective sanitation, reliable energy, and so on provided by engineers. As Peter notes in this book, the

UTAS School of Engineering has produced many very able and some brilliant graduates over the years who have contributed to such infrastructure initiatives.

Further, the esteemed Australian engineer, Sir John Holland in 1988 wrote in his Foreword to 'A Firm Foundation'—the Story of Gutteridge Haskins and Davey that 'One of the fundamental weaknesses in training of young engineers in Australia is an inadequate knowledge of the history of the profession and the achievements of its practitioners'. Peter Doe's treatise would be celebrated by Sir John Holland, as it covers the history and the achievements of UTAS graduates. In addition, it is view of many, including myself that there is an ongoing shortage of local STEM skills and knowledge in Australia, and the book addresses the need to encourage the study of these subjects and thus proceed to enrol in local Australian engineering schools to generate sufficient home-grown engineering capability to drive economic growth in this great nation of ours.

Peter has looked into the future with his final chapter presenting a vision for engineering education. He proposes that design engineers will increasingly rely on simulation and modelling software; however, there is danger in believing the results without knowing and understanding the basis of the models and the simplifying instructions. He stresses that it is vital for graduates to be trained in the fundamental principles of science and engineering and also communication and management skills. He also postulates that Generative AI and virtual and augmented reality tools will have their place, but the same reservations concerning the use of simulation and modelling apply.

Based not only on my engineering practice and education at UTAS, but also on my involvement at the University of NSW, Monash University, and the University of Melbourne engineering schools, I believe that we can all learn much from the study of Peter Doe's book. The book highlights the value of effective, theoretical, practical, and challenging education for engineers in Australia, which is especially pertinent given the significant needs in the Australian engineering and STEM sector at the time of its publication.

Former Board Member and Victoria Manager
Gutteridge Haskins and Davey, Victoria, Australia
Tom Fricke AM

Preface

This book is a labour of love; love for my two sadly departed wives and for the University of Tasmania which has been my life.

My association with the University of Tasmania began in 1947 when my father, the late Thomas Edward Doe, BSc, joined the university staff as a lecturer in education. When I finished high school, I had special permission from the university's Professorial Board to matriculate under-age. I enrolled as a student in the university's Faculty of Engineering in February 1956, seven months before my seventeenth birthday.

In retrospect, it might have been better for me to return to high school for another year. I played in a basketball intervarsity in my first year at the university and generally enjoyed university life; but I failed both pure and applied mathematics. This necessitated repeating most of my first year. I graduated three years later with a Bachelor of Science degree majoring in mathematics and physics. During this time, I joined the University Rugby Club, played in a rugby intervarsity, and was elected to the university's student representative council. I also matured as a student and in the following two years completed a Bachelor of Engineering with Honours Degree majoring in mechanical engineering.

It was at a rugby match where I met my first wife, Cherry Pamela Waters, the sister of David Waters who was playing and daughter of Ken Waters, a keen rugby supporter, graduate of King's College Cambridge, and a member of the university staff in the Department of Classics. Years later, after I had joined the university staff my wife, Cherry, and I were at a university function to greet a newly appointed Vice Chancellor from England. After being introduced to the Vice Chancellor as a lecturer in engineering, I presented my wife as the daughter of Ken Waters and mentioned my father Tom Doe was also on the staff in the Department of Education. My brother-in-law David Waters, who was the University's Acquisition Librarian, joined the conversation. This confirmed what the Vice Chancellor had been told about everybody being related in Tasmania.

Cherry and I were married a year after I joined the Hydro-Electric Commission in Hobart as an engineer in the Hydraulics Section. She continued to work as a primary school teacher until our son, Christopher Thomas Doe, was born in 1968 and two years later our daughter, Alyssa Pamela Doe, after which she became a full-time wife and mother supporting me in my career and raising our children. Later we both returned to high school and university to study the Indonesian language following a period of study leave while I was leading a fisheries research project in Indonesia. Cherry worked as a teacher of Indonesian language until we both retired in 2003. Sadly, after two years travelling around mainland Australia in a caravan as 'grey nomads' she succumbed to cancer. Her legacy is our two successful children. Chris is an airline pilot, a Qantas Captain, with over 18,000 hours of experience. He is married to Maxine, mother of my two adult grandchildren, Logan and Liam. Alyssa graduated in drama but then worked as a first-class flight attendant with Qantas for 23 years while supporting her family. She has now retrained and is working as a counsellor. Her two children, Ellery and Saskia, are year-leaders in high school.

Six months after Cherry died, I was successful in being appointed as Head of Engineering at the Australian College of Kuwait. My time there is chronicled elsewhere in this book, but Kuwait was where I met a lovely lady from Trinidad, Seeta Mohammed, who was working as a teacher in a special needs school. We met at an Arabic Language class. Our friendship blossomed and we were married four years later in 2010. Seeta was an exceptionally talented teacher. She had won awards while working in the USA and had a Master's degree in Special Needs Education from the University of Sheffield. She immigrated to Australia after we were married and, following six months working as an assistant teacher in special education at a school in Hobart, resigned to enrol in a PhD program in the Faculty of Education at the University of Tasmania. She changed her surname to combine her maiden and married names to Seeta Jaikaran-Doe and completed her thesis entitled 'Teachers' Confidence with Technology: Perceptions of the Impact of a Student Laptop Computer Program in Trinidad and Tobago' in three and a half years.

We both took a keen interest in engineering education. At the time I was managing and teaching into a 2 + 2 program in China. Students studied for two years at a Chinese university then moved to the University of Tasmania to complete a Bachelor of Engineering Degree (with Honours) in a further two years. Seeta was able to negotiate a consultancy to develop the pedagogical skills and knowledge of the Chinese teachers. It was when we were both working at the same Chinese university that Seeta sat in on one of my classes and provided feedback which was the basis of the first of our jointly authored technical papers on engineering education. Our contribution to the education of engineering students in China and Australia continued until her untimely death from cancer in 2021.

The University of Tasmania (UTAS) has been my life as an engineer and academic. Even after retiring as an Associate Professor of Engineering in 2003 my involvement continued as the Head of Engineering at the Australian College of Kuwait (ACK) where I set up a program for ACK diplomates to transfer to UTAS to complete an engineering degree. After returning from Kuwait, I have been working at UTAS as a casual lecturer and manager of their Chinese 2+2 program. For two years I returned to full-time work as manager of a major Australian Government's Office of Learning and Teaching project 'Engineering Pathways for Regional Australia'. Then I was engaged full time as a lecturer in the UTAS University College designing and delivering Associate Degree courses. So, it is fitting at a time when my casual teaching appointments in the UTAS School of Engineering appear to have dried up and I am finally retired I have time to author this book.

As a tribute to her contribution to engineering education I have included, as an Appendix, a salute to the life of Dr Seeta Jaikaran-Doe.

Peter Edward Doe

Acknowledgements

This book could not have been written without the contributions from engineers who graduated from the University of Tasmania in the last half of the last century. Recollections of their time as undergraduates are of a time before the ubiquitous laptop and mobile phone. When a reference cited in a lab report, or a technical article was from a physical journal in a library. When graduates had at least some knowledge and understanding of the engineering disciplines at the time—civil, mechanical, and electrical. Their careers as professional engineers span the range of engineering activities. I am grateful for their permissions to publish their 'pitches'. Many thanks to Professor Rupert McLean, AO, who has encouraged me to persevere with writing this book and Tom and Janet Fricke.

Competing Interests The author has no conflicts of interest to declare that are relevant to the content of this book.

Series Editors Introduction

This groundbreaking and insightful book by Peter Doe, *Prospects for Professional Engineering Education in the Asia-Pacific Region: A Historical Case Study of the University of Tasmania,* is the latest volume published in the long-standing Springer Book Series ***Education in the Asia Pacific Region: Issues, Concerns and Prospects***. The first book in this Springer international series was published more than 20 years ago, in 2002, this book by Peter Doe being the 74th volume published to date.

Prospects for Professional Engineering Education in the Asia-Pacific Region examines the history, development, and likely future directions of engineering education in Asia-Pacific countries, during a time of great technological, economic, and social disruption. These areas of disruption, all of which impact on education and schooling, include, but are not limited to: the fourth industrial revolution, including an increasing use of robotics; all-pervasive impact of the internet; Artificial Intelligence; climate/green growth imperatives; and the greening of jobs and work processes. All of these factors impact on the content, teaching modalities, forms of assessment and future directions of education. Peter Doe's book particularly examines the impact of these disrupters on engineering education through a case study of the Faculty of Engineering at the University of Tasmania, Australia.

This detailed case study examines a comprehensive range of cutting-edge issues including (but not limited to): how best to promote the competences of engineering graduates; the increasing impact of STEM on engineering education; changing patterns of student recruitment including more women entering engineering programmes which are increasingly multi-cultural in terms of student recruitment; changes in delivery pedagogy; reengineered visions for the future of engineering education; internationalisation of engineering education programs; the increasing use of simulation and modelling software; and the impact of profound, far-reaching technological change on engineering education competencies.

The author stresses that although it is essential that engineering students are trained in the fundamental principles of science and engineering, it is also important that learners develop well-honed communication and management skills. And Peter Doe refers to the ethical and moral foundations on which he believes engineering education must be built.

As Tom Fricke AM points out in his Foreword, 'this book highlights the value of effective theoretical, practical, and challenging education for engineers in Australia'.

Dr Doe effectively achieves this by focusing in detail on the history, development, and likely future directions of professional engineering education and training in the Faculty of Education in one Australian university, that of the University of Tasmania. But lessons learnt and insights gained are shown to apply to other university schools of engineering throughout Australia and indeed throughout the Asia-Pacific region.

In terms of the Springer Book Series in which this volume is published, the topics dealt with in this book series are wide ranging and varied in coverage, with an emphasis on cutting-edge developments, best practices, and education innovations for development. Topics examined in the series to date include: the interaction between technology and education; the reform of primary, secondary, and teacher education; innovative approaches to education assessment; alternative education; most effective ways to achieve quality and highly relevant education for all; active ageing through active learning; case studies of education and schooling systems in various countries in the region; cross-country and cross-cultural studies of education and schooling; and the sociology of teachers as an occupational group, to mention just a few. More information about the book series is available at http://www.springer.com/series/5888

All volumes in this series aim to meet the interests and priorities of a diverse education audience including researchers, policy makers, and practitioners; tertiary students; teachers at all levels within education systems; and members of the public who are interested in better understanding cutting-edge developments in education and schooling in Asia Pacific.

The main reason why this series has been devoted exclusively to examining various aspects of education and schooling in the Asia-Pacific region is that this is a particularly challenging and dynamic region. It is renowned for its size, diversity, and complexity, whether it be geographical, socio-economic, cultural, political, or developmental. Education and schooling in countries throughout the region impact on every aspect of people's lives, including employment, labour force considerations, education and training, cultural orientation, and attitudes and values. Asia and the Pacific is home to some 63% of the world's population of 7 billion. Countries with the largest populations (China, 1.4 billion; India, 1.3 billion) and the most rapidly growing mega-cities are to be found in the region, as are countries with relatively small populations (Bhutan, 755,000; the island of Niue, in the Pacific, 1600).

Levels of economic and socio-political development vary widely, with some of the richest countries (such as Japan) and some of the poorest countries on earth (such as Bangladesh). Asia contains the largest number of poor of any region in the world, the incidence of those living below the poverty line remaining as high as 40 percent in some countries in Asia. At the same time many countries in Asia are experiencing a period of great economic growth and social development. However, inclusive growth remains elusive, as does growth that is sustainable and does not destroy the quality of the environment. The growing prominence of Asian economies and corporations, together with globalisation and technological innovation, is leading to long-term changes in trade, business, and labour markets, as well as in the sociology of populations within (and between) countries. There is a rebalancing of power, centred on Asia and the Pacific region, with the Asian Development Bank in Manila declaring that the 21st Century will be 'the Century of Asia Pacific'.

We know from comprehensive feedback received from numerous education researchers, policy makers, and practitioners, worldwide, that this book series makes a useful contribution to knowledge sharing about cutting-edge developments concerning education and schooling in Asia Pacific.

Readers of this or other volumes in the series who have an idea for writing or co-writing their own book (or editing/co-editing a book), on any aspect of education and/or schooling, relevant to the region, are enthusiastically encouraged to approach the series editor either direct or through Springer to submit a Publishers Proposal for consideration to publish their own volume in the series. We are always willing to assist in shaping manuscripts in ways that make them suitable for publication.

School of Education
University of Tasmania, Hobart, TAS, Australia
Rupert Maclean AO
Faculty of Education
Monash University, Melbourne, VIC, Australia
Philip K. Chan

19 October, 2024

Contents

About the Author

Peter Edward Doe graduated with degrees in science and engineering from the University of Tasmania in 1963. He worked as an engineer in the Hydraulics Section of the Hydro-Electric Commission in Hobart for two years before accepting a lectureship in mechanical engineering at the University of Tasmania. His industry-funded research led to a patented technique for significantly improving the quality of dried hops. From 1980, he led a team of scientists and engineers, which resulted in world-leading analysis, modelling, and practical methods for improving the quality of sawn hardwood timber.

From 1985 to 1987, he managed a major Australian Centre for International Agricultural Research (ACIAR) program on fish spoilage in Indonesia, which resulted in a significant body of understanding of fish spoilage and the development of a generation of skilled researchers.

In 1996, Dr Doe moved to Launceston to take up the position of Head of the Department of Engineering at the UTAS Newnham campus and for a year was also Head of Architecture. Three years after retiring as Associate Professor of Mechanical Engineering in the UTAS School of Engineering in 2003, he returned to full-time work as Head of Engineering at the Australian College of Kuwait (ACK). After returning to Hobart in 2009 he has continued to lecture part time while managing international engineering articulation programs

Dr Doe is a Fellow of the Institution of Engineers, Australia. He has edited a book (on fish smoking and drying), has written 14 chapters in books, and published more than 80 journal and conference papers including 5 on engineering education.

List of Figures

List of Tables

Chapter 1
Introduction

Abstract This introductory chapter summarizes the book's contents. Starting with a brief history of engineering, there follow descriptions of the engineering courses in each of the Asia-Pacific countries which are signatories to the Washington Accord which resulted from a meeting of national professional engineering accreditation bodies. The history and development of the Bachelor of Engineering degree at the University of Tasmania is presented as a case study. Included are personal histories and undergraduate experiences of the engineering students at the University of Tasmania in the 1960s, 1970s, and 1980s. Their successful careers as engineers and managers are a testament to the quality of their engineering education. The following chapters record the contribution to engineering education from the collaboration of this author and Dr. Seeta Jaikaran-Doe. The book concludes with the author's vision for the future of engineering education.

This book has resulted from my passion to publish parts of Dr. Seeta Jaikaran-Doe's PhD thesis together with the many personal histories and undergraduate experiences of the engineering students I lectured to at the University of Tasmania in the 1960s, 1970s, and 1980s. Their successful careers as engineers and managers are a testament to the quality of their engineering education. Building on this base, the book presents a wider view of professional engineering education in Australia's Asia-Pacific neighbors, those countries who are signatories to the "Washington Accord" for worldwide accreditation of engineering courses at higher education institutions. It is a personal view resulting from my lifelong career as an engineering academic with the University of Tasmania.

The chapter following this introduction starts with a brief history of engineering from the time of the pyramids in Egypt. The word "engineer" is applied to many occupations from train drivers to computer programmers. This book is confined to the training of professional engineers who have completed a 3- or 4-year university degree. What follows in this chapter is a review of the training of professional engineers in universities in the Asia-Pacific region.

The third chapter focuses on the training of professional engineers in Australia drawing on my experience at the University of Tasmania as a case study. Details of

P. E. Doe, *Prospects for Professional Engineering Education in the Asia-Pacific Region*, Education in the Asia-Pacific Region: Issues, Concerns and Prospects 74,
https://doi.org/10.1007/978-981-96-3197-1_1

the curriculum and its relevance to the competence of engineering graduates are explored.

The fourth chapter chronicles the history of the Faculty of Engineering at the University of Tasmania. It recounts some of the twists and turns and the role of personalities and ambitions in evolving from a "Faculty of Engineering" to a combined "Department of Engineering and ICT" and the "divorce" when ICT (Information and Communications Technology) emerged as a separate department. It is a personal account because I was the head of the Department of Engineering in Tasmania's second city, Launceston, between 1996 and 2000 when the review of engineering education in Tasmania recommended the transfer of the first year "feeder course" into the Australian Maritime College and the teach-out of my department's environmental engineering BEngTech course.

Chapter 5 includes presentations by UTAS—trained professional engineers, many of whom I taught, who graduated in the 1940s, 1950s, 1960s, 1970s, and later. These presentations were made at an online session in October 2021 held to celebrate the 100th anniversary of the first meeting of the UTAS Faculty of Engineering. They are included, verbatim, and reflect on their time as undergraduates and the variety and eminence of their subsequent careers. I have also included biographies and "pitches" from UTAS Rhodes Scholars.

Chapter 6 looks at the period of internationalization in engineering education which saw Australian and New Zealand universities actively recruiting international students with universities sending staff to neighboring countries in the Asia-Pacific region to deliver engineering courses and set up campuses.

Chapter 7 is about the pedagogy of engineering education based on the collaboration between this author and Dr. Seeta Jaikaran-Doe. Following a survey of engineering students' engagement with their lecturers' presentations and online material conducted by a final year student, we turned our attention to the Technological Pedagogical Content Knowledge (TPACK) competency of the UTAS School of Engineering staff and then followed work on strategies for enhancing student learning in an Australian 2 + 2 undergraduate engineering program and an investigation into the articulation of the 2 + 2 Chinese students into the UTAS engineering degree programs. The chapter concludes with some of the findings of a review of engineering lecturers' delivery practices post-COVID.

In keeping with the original motivation for publishing this book—that is, to publish the part of Dr. Seeta Jaikaran-Doe's PhD thesis—I have included in this book as Chap. 8 her framework for quality teaching and learning in the twenty-first century. It was developed from extensive research into the TPACK of high school teachers in Trinidad and Tobago, but, as reported in Chapter 7, it is directly applicable to learning and teaching at university level.

Chapter 9 is my vision for the future of engineering education. If one looks back at how much engineering and engineering education has changed over the past 60 years post-computers, it would be brave or foolish of me to be confident in my predictions.

As a tribute to her contribution to engineering education, I have included, as an appendix, a salute to the life of Dr. Seeta Jaikaran-Doe.

Chapter 2
Engineering Education in the Asia-Pacific Region

Abstract This chapter starts with a brief history of engineering from the time of the pyramids in Egypt. The word "engineer" is applied to many occupations from train driver to computer programmers. The industrial revolution in the mid-eighteenth-century saw the need for training of engineers who understood the scientific and industrial principles of structures, mechanics, heat engines, electric motors, and fluid mechanics. Higher education institutions were founded to train what are now known as professional engineers who have completed a 3- or 4- year university degree. Then follows a review of professional engineering education in each of the Asia-Pacific signatories to the Washington Accord: Australia, China, Hong Kong, India, Indonesia, Japan, Korea, Malaysia, New Zealand, Pakistan, Singapore, Sri Lanka, and Taiwan. The history, status, and curricula of undergraduate professional engineering courses in each of these countries are examined.

2.1 Introduction

This chapter starts with a brief history of the profession of engineering and follows the development of engineering education as it evolved from the master-apprentice model in ancient times to the establishment of faculties of engineering in universities throughout the Asia-Pacific region and the signing of the Washington Accord which regulates the training of today's professional engineers. The evolution of professional engineering education at the University of Tasmania is presented as a case study in Chap. 4 from the perspective of the author of this book who has been associated with the University of Tasmania since 1947 and its Faculty of Engineering since enrolling as an undergraduate in 1957.

P. E. Doe, *Prospects for Professional Engineering Education in the Asia-Pacific Region*, Education in the Asia-Pacific Region: Issues, Concerns and Prospects 74,
https://doi.org/10.1007/978-981-96-3197-1_2

2.2 Engineering History

Civil engineering dates to the earliest time of civilization. The word "civil" distinguishes what we now know as civil engineering today from "military" engineering in ancient times which was to do with fortifications, siege-breaking machines, moats, and the like. The first engineer in recorded history, Imhotep, built the pyramids in Egypt.

Chinese engineers under the direction of Qin Shi Huang, the first emperor, began to build the Great Wall of China as a defense strategy in the seventh century BC.

The study of mathematics, chemistry, astrology, and medicine flourished in the Arab world while Europe was in the "dark ages." The engineers Anthemius of Tralles and Isidorus the Elder are credited with the construction of the Hagia Sophia in Istanbul in the sixth century. Ismail al-Jazari described 100 "mechanical devices" in the thirteenth century.

Civil engineering and military engineering preceded the era of mechanical and electrical engineering. Leonardo da Vinci, James Watt, Isambard Kingdom Brunel, Gustave Eiffel, Nikola Tesla, Thomas Edison, Henry Ford, Alexander Graham Bell, and Nikolaus Otto are names of celebrated civil, mechanical, and electrical engineers. In their time, there were very few places where bright young men and women could receive training as engineers. Mostly they learned from their predecessors; many learned from their mistakes.

The Ecole Polytechnique was founded in 1794 to deal with a shortage of engineers in France. Students were given a grounding in physics, chemistry, and mathematics. The Technische Universitat and Hochschule were established throughout Germany at about the same time.

These and similar colleges in Europe and elsewhere differed from the existing universities which traditionally enshrined and progressed knowledge of the arts, philosophy, medicine, and law.

The Corps of Engineers was founded in the UK in 1717 with the Institution of Civil Engineers following a century later.

At the beginning of the nineteenth century, there was an increasing need for engineers to build roads and bridges, design tall buildings, and develop electrical power and communication systems. Undertaking these tasks necessitated an understanding of basic principles of physics, mathematics, chemistry, and geology together with the knowledge and skills to apply codes of practice. In addition, engineers needed to be skilled in communication and management.

Malcolm Spencer Gregory was appointed to a lectureship at the University of Tasmania in 1956. As well as pioneering research into the analysis of structures, he published a book on the history of engineering (Gregory, 1971). Malcolm Gregory was the first person to be awarded the degree of Doctor of Philosophy in the Faculty of Engineering at the University of Tasmania.

In the preface to his book on the history of engineering, Malcolm Gregory writes:

> Engineering design is an art based on knowledge of many kinds and the ability to conceive and carry through a project to completion. Creative design involves working at the frontier

> of knowledge and pushing into the unknown in order to achieve the aims of the designer. An engineer uses as much formalised science as is available to aid him but relies for the greater part of his design planning and execution of works on that great body of much more incoherent knowledge which lies at the base of all human action. The designer plans work to be carried out by and for men and is dependent for his resource is on all aspects and institutions of organised society. (Gregory, 1971. p. vii)

Gregory also quotes Vitruvius on the training of "architects" as civil engineers were known in Roman times:

> The architect must both have an actual gift and be ready to learn. Neither talent without instruction nor instruction without talent can produce the perfect craftsman. The architect should be a man of letters to keep a record of useful precedence; a skilful draughtsman who can represent by coloured drawings of the effect desired; a mathematician who can use rule and compass, lay out works, use optics to consider the effect of light and arithmetic to add up costs; a historian to give account of historical events behind his decorations; a diligent student of philosophy to make him high minded, not arrogant, fair-minded without avarice, also understanding the principles of nature and the flow of water; not ignorant of medicine for works of water supply and drainage; learned in the law for contract specifications and disputes; and finally familiar with astronomy at astronomical calculations. Training should be from boyhood a general education is put together as one body from its members. (Gregory, 1971. p. 17)

Engineers, particularly civil, electrical power, and aeronautical engineers, have a responsibility to the public. Errors they make can have fatal consequences. In many countries, experienced engineers formed bodies such as the Institution of Engineers, Australia (established in 1919), to assess and accredit universities and other bodies training professional engineers.

2.3 Asia-Pacific Signatories to the Washington Accord

Representatives from accrediting bodies in the UK, USA, Canada, Australia, Ireland, and New Zealand met in 1989 to sign the Washington Accord which sets the standard for training engineers at universities in those countries. Engineers certified under the Washington Accord must have completed a 3- or 4-year university-level degree following 12 or 13 years of school education. The degree course must include elements of mathematics, physics, and chemistry together with an in-depth understanding of engineering principles, design, and practice. There must be a management component and a commitment to ethical practices. Graduates must be able to communicate in lay and technical language. In 2024, there are 23 signatories to the Washington Accord: Australia, Canada, China, Cost Rica, Hong Kong, India, Indonesia, Ireland, Japan, Korea, Malaysia, Mexico, New Zealand, Pakistan, Peru, Russia, Singapore, South Africa, Sri Lanka, Taiwan, Turkey, the UK, and the USA. Graduates from accredited universities in those countries can call themselves professional engineers and are eligible to be employed as such in any of the signatory countries.

In 2021, the International Engineering Alliance, supported by UNESCO and the World Federation of Engineering Organizations, published revisions to its Graduate Attributes and Professional Competencies international benchmark. This document sets forward the attributes and competencies for the three levels of engineering roles—engineering technicians (Dublin Accord), engineering technologists (Sydney Accord), and professional engineers (Washington Accord) (IEA, 2021).

2.3.1 Washington Accord Graduate Attributes (GAs)

A professional engineer should be able to:

- Comprehend and apply advanced knowledge of the widely applied principles underpinning good practice.
- Comprehend and apply advanced knowledge of the widely applied principles underpinning good practice specific to the jurisdiction of practice.
- Define, investigate, and analyze complex problems using data and information technologies where applicable.
- Design or develop solutions to complex problems considering a variety of perspectives and taking account of stakeholder views.
- Evaluate the outcomes and impacts of complex activities.
- Recognize the foreseeable economic, social, and environmental effects of complex activities and seek to achieve sustainable outcomes (as represented by the 17 UN Sustainable Development Goals).
- Meet all legal, regulatory, and cultural requirements and protect public health and safety in the course of all activities.
- Conduct activities ethically.
- Manage part or all of one or more complex activities.
- Communicate and collaborate using multiple media clearly and inclusively with a broad range of stakeholders in the course of all activities.
- Undertake CPD activities to maintain and extend competences and enhance the ability to adapt to emerging technologies and the ever-changing nature of work.
- Recognize complexity and assess alternatives in light of competing requirements and incomplete knowledge. Exercise sound judgment in the course of all complex activities.
- Be responsible for making decisions on part or all of complex activities.

2.3.2 Washington Accord Course Structure

A Washington Accord program provides:

- A systematic, theory-based understanding of the natural sciences applicable to the discipline and awareness of relevant social sciences.

- Conceptually based mathematics, numerical analysis, data analysis, statistics, and formal aspects of computer and information science to support detailed analysis and modeling applicable to the discipline.
- A systematic, theory-based formulation of engineering fundamentals required in the engineering discipline.
- Engineering specialist knowledge that provides theoretical frameworks and bodies of knowledge for the accepted practice area in the engineering discipline; much is at the forefront of the discipline.
- Knowledge, including efficient resource use, environmental impacts, whole-life cost, reuse of resources, net-zero carbon, and similar concepts, that supports engineering design and operations in a practice area.
- Knowledge of engineering practice (technology) in the practice areas in the engineering discipline.
- Knowledge of the role of engineering in society and identified issues in engineering practice in the discipline, such as the professional responsibility of an engineer to public safety and sustainable development.
- Engagement with selected knowledge in the present research literature of the discipline, awareness of the power of critical thinking, and creative approaches to evaluate emerging issues.
- Ethics, inclusive behavior, and conduct. Knowledge of professional ethics, responsibilities, and norms of engineering practice. Awareness of the need for diversity by reason of ethnicity, gender, age, physical ability, etc., with mutual understanding and respect and of inclusive attitudes.

A program that builds this type of knowledge and attitude and develops the Graduate Attributes listed above is typically achieved in 3–5 years of study, depending on the level of students at entry.

What follow in this chapter are synopses of the professional engineering programs in all the signatory countries in the Asia-Pacific region, their history, accreditation protocols, the attributes of their graduates, and the course structure assuring those attributes.

2.3.3 Australia

2.3.3.1 History

The University of Sydney was established in 1850, followed by the University of Melbourne (1853), the University of Adelaide (1874), and the University of Tasmania in 1890. Teaching of engineering at the University of Sydney started in 1883 with the other Australian universities appointing engineers as professors and lecturers, mostly sourced from the established UK universities in the decades that followed. The Faculty of Engineering at the University of Sydney was not established until 1920, only 2 years before that of the University of Tasmania which is the case study in this book.

2.3.3.2 Accreditation

Since the establishment of the Institution of Engineers, Australia (IEAust), in 1919, all professional engineering courses in Australia have been scrutinized and accredited by a panel of professional engineers appointed by the Institution. The process involves the degree-awarding higher education authority (HEA) providing extensive documentation to the panel which, after reviewing the documentation, visits the HEA to meet with staff and students and inspect facilities. Accreditation can be awarded for a 3- or 5-year period. It is normal for the accreditation panel to require the HEA to rectify any perceived deficiencies in a course or its delivery. Full accreditation is awarded only after students have graduated, at which time their graduate attributes can be assessed.

2.3.3.3 Course Structure

There is no common course structure in Australia for the graduation of professional engineers. The most common structure is 4 years with an entry after school year 12 with direct entry students having studied the highest level of mathematics, physics, and chemistry. Some universities provide bridging courses, and there are provisions for mature age entry with appropriate engineering experience or TAFE (Technical and Further Education) qualifications. Swinburne University has traditionally offered a 5-year professional engineering degree which includes 12 months of paid industry placement. The University of Melbourne awards a (nonprofessional) Bachelor of Engineering degree after 3 years full-time study with a further 2 years to gain a professional Master of Engineering degree. Master of Professional Engineering degrees are becoming a popular choice for students to study in Australia who have graduated with degrees from HEAs in other countries which are not accredited under the Washington Accord.

2.3.4 China

2.3.4.1 History

Chinese architects, engineers, and scientists are credited with the invention of the abacus, the sundial, magnetic compass, gunpowder, papermaking, and printing. The Tang dynasty (AD 618–906) is recognized as a time of great invention. The Great Wall of China is one of the Seven Wonders of the Ancient World.

Until the mid-1880s, there was no formal training of engineers in China. The first academies to train engineers were established as a result of the humiliating defeat of China's army and navy by the British and Japanese in the Opium and Sino-Japanese wars. At the time engineers were in the third rank of Chinese society, below the farmers in the second rank. The highest rank was that of a civil servant which was open only to those who passed the Civil Service Examination based largely on Chinese language, history, and the Confucian philosophy.

The Foochow Shipbuilding College was founded in 1866 to design and build a modern naval fleet (McNeill et al., 2010). There were two streams—one to train engineering students in the design and construction of ships and components and the other to prepare naval officers. The engineering students followed a French curriculum based on that of the Ecole Nationale des Ponts et Chaussées which had been established 120 years earlier. The naval officers were trained under the British system which had the reputation of producing the world's best navigators. The French school's textbooks were in the French language, and students studied arithmetic, trigonometry, analytic geometry, calculus, physics, and mechanics in their first 2 years before choosing "engineering cartography" or "the manufacture and structure of steam engines" or "manufacturing of ship hulls" as majors. The British school's basic sciences included arithmetic, geometry, calculus, and spherical trigonometry with majors in naval astronomy, meteorology, and geography.

In 1895, the Peiyang University was established in Tianjin as the first Western-style institution for training engineers. It differed from other Chinese higher education institutions; the delivery style was learner-centered with an emphasis on practical applications and creative abilities (McNeill et al., 2010). Professors were recruited from the West with scientific and laboratory equipment purchased from the USA. The university also subscribed to more than 100 Western scientific and engineering journals. Civil, metallurgical, and mining engineering departments were established and later a department of railway engineering. Notably, students were able to decide on their specialization at any time in the first or second years of their 4-year course.

The Imperial Railway College (est. 1896) is recognized as the "birthplace of China's modern education in civil engineering" (SWJTU, n.d.). It was established with mainly foreign academics to train engineers to design and build China's railway system. The Nanyang College in Shanghai (now named Shanghai Jiao Tong University) was also founded in 1896 and followed the practical, Western-style approach. Civil, mechanical, and electrical specializations were introduced in 1905.

In June 2016, China officially became a full member of the Washington Accord.

There are now 2893 engineering degrees accredited by the China Engineering Education Accreditation Association (CEEAA) to deliver professional engineering courses in China under the Washington Accord.

Prashant Loyalka, Stanford University, contributed an appendix to World Bank report (World Bank, 2013). Parts of this appendix which are directly relevant to the topic of this chapter are reproduced below (readers should refer to the original for citations which are not included here):

Background

Modern engineering was introduced in China in the late 19th century after a series of military defeats convinced the Qing government to modernize the military. However, engineering education did not become systematic until after the establishment of China's first modern universities at the turn of the 20th century. After 1949, engineering education was reorganized when the government transformed the country's early universities into smaller, highly specialized institutions created to meet the demands of industrialization. These Soviet-style institutions grew rapidly, and enrollment increased nearly sixfold between 1949 and 1965, peaking at around one million in 1960. This expansion persisted until the launch of the Cultural Revolution in 1966, at which point many higher education institutions were closed, and academia stagnated for the next 10 years.

Beginning in the 1970s, economic reforms such as the Four Modernizations emphasized science, technology, and engineering as key to China's economic modernization and development. As a result, the proportion of engineering students in universities rose, reaching a third of total enrollments and lingering between 35 and 40 percent throughout the 1980s and the 1990s. By 1998, at the start of the government's mass effort at the expansion of higher education, the number of four-year (undergraduate) engineering students was around 1.1 million. The number increased by almost 3.5 times over the next decade and, by 2010, had reached close to 4.0 million.

Issues and challenges

China still faces many challenges in guaranteeing the quality of engineering education, especially in non-elite institutions at the undergraduate level and at the graduate level in general. One issue is that the quality of courses, curricula, and instruction may be substandard in the less- selective non-elite institutions. While the Ministry of Education sets the total number of credit hours per major and designates certain courses that should be studied, course specifics, including the number of credit hours devoted to particular types of courses, the choice of curricular materials, and instructional practices, is left to the discretion of individual institutions and faculty.

Another issue is that students still seem to lack practical (research and internship) experience, as well as exposure to certain soft skills that are emphasized in the U.S. model of engineering education. For example, only a small share of engineering students appear to be participating in faculty research projects, engaging in meaningful internships, or taking leadership or interdisciplinary courses. Furthermore, four-year students are generally known to complete a rather concentrated, classroom-oriented course schedule in the first three years, but then spend much of their last year searching for jobs.

Further impacting the quality of engineering graduates is the fact that students in China (especially in the less-selective institutions) have few incentives to perform well in college because they are, essentially, guaranteed a diploma after the designated period of study. This raises questions about the relative value added in Chinese engineering programs. If Chinese students are guaranteed graduation, this may reduce student learning, particularly at nonelite institutions. More specifically, while students at the best Chinese universities and at the more selective second-tier institutions may be motivated to study seriously in college to enter graduate school at home or abroad, the majority of students in the less selective second-tier universities and those in third-tier universities may have much less incentive to perform well in their studies.

Finally, it is an outstanding question whether Chinese engineering graduates meet employer demands. According to somewhat dated reports from the mid-2000s, employers find that the overall quality of Chinese engineering graduates is rather low. Specific problems include a mismatch between the skills and knowledge possessed by graduates and those required by companies, a lack of practical engineering experience, an absence of creativity and risk-taking behavior, and substandard English and communication skills. At the same time, however, the employment rate among recent engineering graduates is fairly high (around 85–90 percent), as is the rate of returns to college, especially in comparison with graduates in other majors. Retrieved from: https://hdl.handle.net/10986/16562

2.3.4.2 Accreditation

The China Engineering Education Accreditation Association (CEEAA) was established in October 2015. It is responsible for the organization and implementation of engineering education accreditation in China. CEEAA is responsible for forming a quality assurance system for engineering education, establishing an engineering education accreditation system linking up with the professional engineering system, promoting the cooperation between engineering education and industry, and promoting the international mutual recognition of Chinese engineering education and enhancing communications with international counterparts.

2.3.4.3 Graduate Attributes

The listed attributes of graduates in China (CEEAA, 2022) are similar to those promulgated by the Washington Accord (IEA, 2021), namely:

- Engineering knowledge: Apply knowledge of mathematics, natural science, engineering fundamentals, and engineering specialization to solve complex engineering problems.
- Problem analysis: Identify, formulate, research literature, and analyze complex engineering problems reaching substantiated conclusions using basic principles of mathematics, natural sciences, and engineering sciences.
- Design/development of solutions: Design solutions for complex engineering problems and design systems, components, or processes that meet specified needs with appropriate societal, public health, and safety, legal, cultural, and environmental considerations.
- Investigation: Conduct investigations of complex problems using research-based knowledge and research methods, including design of experiments, analysis and interpretation of data, and synthesis of the information to provide valid conclusions.
- Modern tool usage: Create, select, and apply appropriate techniques, resources, modern engineering, and IT tools for complex engineering problems, including prediction and modeling of complex engineering problems, with an understanding of the limitations.
- Engineer and society: Apply reasoning informed by the contextual knowledge to assess societal, health, safety, legal, and cultural issues and the consequent responsibilities relevant to professional engineering practice and solutions to complex engineering problems.
- Environment and sustainability: Understand and evaluate the sustainability and impact of professional engineering work in solving complex engineering problems in societal and environmental contexts.
- Professional ethics: Have humanities and social science qualities and social responsibility, apply ethical principles, and commit to professional ethics and responsibilities and norms of engineering practice.

- Individual and teamwork: Function effectively as an individual, team member, and principal in a multidisciplinary team.
- Communication: Communicate effectively on complex engineering activities with the engineering community and with society at large, such as being able to comprehend and write effective reports and design documentation, make effective presentations, and give and receive clear instructions. Have a particular international perspective, communicate, and exchange in the cross-cultural context.
- Project management: Understand and master engineering management principles and economic decision-making methods, and apply them in a multidisciplinary environment.
- Lifelong learning: Recognize the need for and have the preparation and ability to engage in independent and lifelong learning in the broadest context of technological change.

2.3.4.4 Course Structure

The CEEAA (2022) also mandates the following generic course structure:

- The curriculum must be consistent with graduate outcomes. The design of the curriculum must involve experts from the enterprises or industry. The curriculum must include:
- Courses on mathematics and nature sciences consistent with the graduate outcomes (accounting for at least 15% of the total credits).
- Courses on engineering fundamentals, courses on subject fundamentals, and subject courses (accounting for 30% of the total credits). Courses on engineering fundamentals and courses on subject fundamentals may provide training in the ability to apply mathematics and natural science in solving complex problems related to the professional discipline. Subject courses can fully assume the role of training abilities in system design and operation.
- Engineering practice and graduate design (thesis) (accounting for 20% of the total credits). The program has a well-established practice education system and cooperates with enterprises to educate students on practical and innovative abilities. The topics of graduate design (thesis) are oriented from the practical engineering problem to educate students on engineering awareness, cooperation, and abilities to systematically utilize what they have learned to solve complex engineering problems. The guidance and evaluation of graduation design (thesis) involve experts from industry or enterprises.
- Courses on humanities, social sciences, and general education (accounting for at least 15% of the total credits) to enable students to consider the economic, environmental, legal, safety, health, and ethical constraints in engineering practice.

2.3.5 *Hong Kong*

2.3.5.1 History

The University of Hong Kong was founded in 1911 as a "British University on Chinese Soil" (Wu, 1992). Presently, there are 11 universities offering engineering degrees in Hong Kong with several ranked highly worldwide. The Hong Kong Polytechnic University (HKPU) is the top-ranked university in Hong Kong and was established 85 years ago specifically for training engineers. HKPU claims to have pioneered work-integrated programs in their undergraduate curriculum.

2.3.5.2 Accreditation

Engineering degrees in Hong Kong at all levels from professional to technician are assessed and accredited by the Hong Kong Institution of Engineers. Assessment recognizes the importance of the quality of the engineering graduates as distinct from the curriculum:

> The quality of an engineering degree programme depends on more than just the curriculum and syllabus. The quality of the graduates is an important consideration in the evaluation of an engineering programme. The degree programme must define outcomes that they expect of their graduates consistent with their educational objectives and the needs of the discipline; and describe the processes that are used to measure and evaluate these outcomes. In addition, the calibre of the academic staff, the entry standards, staffing levels, teaching methods, facilities, funding and method of assessment are just some of the factors which influence the quality of the educational experience and the outcomes. (HKIE, 2013)

2.3.5.3 Graduate Attributes

Engineering programs accredited by the HKIE must demonstrate that their graduates have the following attributes (HKIE, 2013):

- An ability to apply knowledge of mathematics, science, and engineering appropriate to the degree discipline
- An ability to design and conduct experiments, as well as to analyze and interpret data
- An ability to design a system, component, or process to meet desired needs within realistic constraints, such as economic, environmental, social, political, ethical, health and safety, manufacturability, and sustainability
- An ability to function on multidisciplinary teams
- An ability to identify, formulate, and solve engineering problems
- An ability to understand professional and ethical responsibility
- An ability to communicate effectively

- An ability to understand the impact of engineering solutions in a global and societal context, especially the importance of health, safety, and environmental considerations to both workers and the general public
- An ability to stay abreast of contemporary issues
- An ability to recognize the need for and to engage in lifelong learning
- An ability to use the techniques, skills, and modern engineering tools necessary for engineering practice appropriate to the degree discipline
- An ability to use the computer/IT tools relevant to the discipline along with an understanding of their processes and limitations

2.3.6 India

2.3.6.1 History

The Engineering College at Roorkee in the state of Uttarakhand, established in 1847, claims to be the first institution in India to award engineering degrees. It was followed in 1854 by the Poona Civil Engineering College (1854) and the Bengal Engineering College at Shibpur in 1865. These colleges followed the British model before Indian independence but changed to the US model after independence in 1947 (Chinna Balu, 2014). Before Indian independence, there were only 35 engineering colleges in India. By 1975 this number had grown to 437. In 2021 there were 2500 engineering colleges and 1400 polytechnics.

In 1944 a committee chaired by NR Sarkar was tasked to "consider the development of Higher Technical Institutions in India with a view to ensuring an adequate supply of technical personnel which will be required for post–War industrial development in this country." The Sarkar Committee recommended higher technological institutes based on the Massachusetts Institute of Technology. Five Indian Institutes of Technology (IITs) were established at Bombay (1958), Kharagpur (1959), Kanpur (1959), Madras (1960), and Delhi (1961).

The World Bank undertook a study in 2013 to compare engineering education in India with China, Japan, Brazil, Russia, the USA, and the UK (World Bank, 2013). The study found India to have the lowest number of engineering students per million of population (Fig. 2.1).

The World Bank's remit was to compare numbers of students and faculty in the institutions together with expenditure and education, expenditure, research, and commercialization outputs. Their remit did not include curricula, employability, pedagogy, learning outcomes, and assessments. The Word Bank noted there were a large number of higher education institutions in India, but access to engineering education remains an issue; there is a substantial shortage of qualified faculty; only limited data were available on the engineering education system. They did note, however, the quality and number of research activities had increased, but there was room for more improvement.

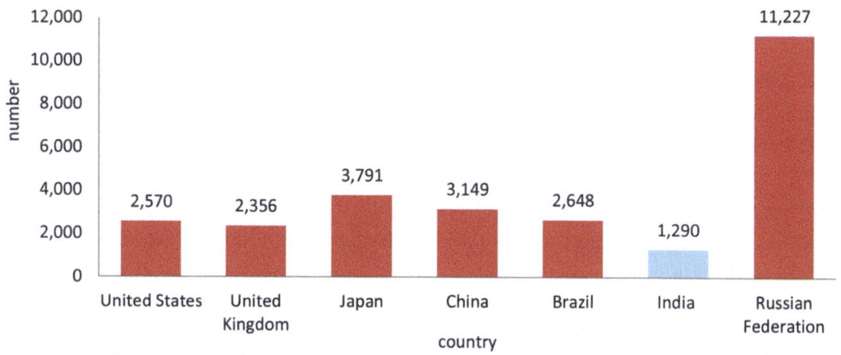

Sources: Various (see the data sources described in appendix E); population data: World Bank 2010.
Note: Students include undergraduate and postgraduate students.

Fig. 2.1 Number of engineering students, per million population (World Bank, 2013)

2.3.6.2 Accreditation

Indian engineers are recognized as professional engineers under the Washington Accord (WA) if they have graduated from one of the tier 1 education institutions recognized by the National Board of Accreditation (NBA). The NBA website (https://www.nbaind.org/Accreditation) lists 1845 accredited degrees. In recent years, Australian universities have introduced 2- or 3-year master's programs for international students wanting to upgrade non-WA-accredited degree to qualify for graduate membership of Engineers Australia. Experience at the University of Tasmania is students who are enrolled in the third- and fourth-year units which are part of the UTAS Master of Professional Engineering degree program experience difficulties through not having undertaken the underpinning maths, science, and engineering units from first and second year.

2.3.6.3 Graduate Attributes

Graduate attributes are mandated by the NBA. They are:

- Engineering knowledge
- Problem analysis
- Design/development of solutions
- Conduct investigations of complex problems
- Modern tool usage
- The engineer and society
- Environment and sustainability
- Ethics
- Individual and Team Work10
- Communication
- Project management and finance
- Lifelong learning

2.3.6.4 Course Structure

In 2009, NBA aligned its methodology with international benchmarks and started accreditation on the basis of outcomes. It believes that educational quality must be measured by outcomes rather than inputs, because inputs do not necessarily correlate with quality outcomes. Outcomes are dependent not only on inputs but also on the processes followed by an institution to convert inputs into defined outcomes. Accordingly, the NBA accreditation process does not mandate course structure.

2.3.7 Indonesia

2.3.7.1 History

The first higher education institution to deliver engineering courses in Indonesia, *De Technische Hoogeschool te Bandung (Technical High School in Bandung*), was established in 1920 during the Dutch colonial time. The first major was in the Road and Hydraulic Engineering Faculty (*de afdeeling der We gen Waterbouw).* In the period before and after Japanese occupation and independence in 1947, engineering colleges were established throughout Indonesia. In 2023 there were 65 universities in Indonesia offering 104 engineering degrees accredited by the Indonesian Accreditation Board for Engineering Education (IABEE) and recognized internationally under the Washington Accord. The three highest Southeastern Asia universities listed in the 2024 QS Asia University Rankings are Universitas Indonesia (10), Gadjah Mada University (12), and Bandung Institute of Technology (13). Indonesian higher education establishments are mostly ranked lower than their counterparts in Singapore and Malaysia.

2.3.7.2 Accreditation

The Indonesian Accreditation Board for Engineering Education applies the following set of outcome-based criteria for evaluating engineering degree programs (IABEE, 2024). Note that individual degree programs are assessed and not their host universities although the provision of resources such as laboratories, computers, and quality teaching and research faculty is generally university-wide responsibilities:

2.3.7.3 Graduate Attributes

The IABEE mandates a professional engineering program shall define the profile of graduates to be envisaged as autonomous professionals by considering country's potential resources, cultures, needs, and interests. The program shall establish its expected learning outcomes which consist of abilities to utilize knowledge, skills, resources, and attitudes as described in the following items to be acquired by the student at the time of completion of the study:

- An ability to apply knowledge of mathematics, natural and/or materials sciences, information technology, and engineering to acquire comprehensive understanding of engineering principles
- An ability to design components, systems, and/or processes to meet desired needs within realistic constraints in such aspects as law, economic, environment, social, politics, health and safety, and sustainability as well as to recognize and/or utilize the potential of local and national resources with global perspective
- An ability to design and conduct laboratory and/or field experiments as well as to analyze and interpret data to strengthen the engineering judgment
- An ability to identify, formulate, analyze, and solve complex engineering problems
- An ability to apply methods, skills, and modern engineering tools necessary for engineering practices
- An ability to communicate effectively in oral and written manners
- An ability to plan, accomplish, and evaluate tasks under given constraints
- An ability to work in multidisciplinary and multicultural team
- An ability to be accountable and responsible to the society and adhere to professional ethics in solving engineering problems
- An ability to understand the need for lifelong learning, including access to the relevant knowledge of contemporary issues

2.3.7.4 Course Structure

The curriculum shall include the following subject areas:

- Mathematics and discipline-specific natural sciences
- Discipline-specific engineering science and technology
- Information and communication technology
- Engineering design and problem-based experiments
- General education, which includes morality, ethics, socio-culture, environment, and management (IABEE, 2024)

2.3.8 Japan

Tsutomu Kimura, National Institution for Academic Degrees, provided an appendix to the World Bank report entitled: Overview of higher education and the engineering education system. Parts of this appendix which are directly relevant to the topic of this chapter are reproduced below:

2.3.8.1 History

The promotion of industry was the first priority of the new government created immediately after the Meiji Restoration (1868). This required high-quality engineering education. Because no Japanese instructors were available, the prime minister contacted personal acquaintances in the UK to seek a proper faculty. His request reached William John Macquorn Rankine, father of modern thermodynamics, at Glasgow University. Rankine recommended one of his students, Henry Dyer, to the Japanese government. Dyer was an extremely bright engineer and only 25 years of age.

In those days, there was no academic discipline called engineering in the UK. There was only the practical application. Because Dyer did not believe this was an appropriate approach, he decided to found an engineering school in Japan. There, he integrated the practice-oriented British style with the highly academic European style. The engineering school was opened in 1873 as the Imperial College of Engineering. Dyer started a 6-year course of study at the school according to the following structure:

- Years 1 and 2: mainly classroom teaching
- Years 3 and 4: 50 percent experiments and practical experience
- Years 5 and 6: full internship at state-run factories or other jobsites

- The school was enormously successful, and it produced many innovative engineers who contributed to the modernization of Japan. It was greatly influenced by British engineering education. It remained independent for only 12 years, becoming a part of Tokyo University in 1885. Responsibility for the school was shifted from the Ministry of Trade and Industry to the Ministry of Education. This ministry altered the school's system: coursework was reduced to 4 years, and the practical elements in the curricula were cut substantially. As a result, engineering as an academic discipline produced fewer and fewer high-quality engineers. This trend is still evident today.

2.3.8.2 Accreditation

In 2005, the Japan Accreditation Board for Engineering Education became a signatory member of the Washington Accord, an international accreditation agreement on professional engineering academic degrees. On that occasion, the review team commented that most Japanese engineering programs emphasize the learning of relevant scientific principles more than the application of those principles in a design context, making Japanese engineering education somewhat different from that found in many of the countries of the Washington Accord. The review team suggested that this meant that, although the end result is clearly highly educated engineering graduates with excellent experience in research, the Japanese graduates probably have little hands-on engineering experience.

In 2012, when new recognition criteria were adopted, the accreditation board began placing a greater focus on the design ability of engineering graduates.

Japanese engineering education at the tertiary level comes under the purview of the general higher education system. The higher education system is composed of national, private, and prefectural universities. Undergraduate courses last 4 years, and the master's degree usually takes 2 years. PhD degrees can normally be obtained with an additional 3 years of study after the master's. The participation ratio in higher education is 52%, and the completion rate is 92%.

There are 86 national, 589 private, and 90 prefectural universities. The total number of students is 2,835,000, among whom 624,000, 2,080,000, and 132,000 are enrolled at national, private, and prefectural universities, respectively. Around 2,520,000 students are in undergraduate programs, 165,000 in master's programs, 74,000 in PhD programs, and 23,000 in professional programs. Women students account for 40%, 30%, 31%, and 28% of the students in these respective courses. These proportions have been steadily increasing in recent years. In the field of engineering, however, the proportion of women students has remained unchanged, at only 11%.

In terms of student enrollments, social science has the largest share, at 33% (928,000 students), whereas engineering represents 17% (489,000 students).

The number of part-time undergraduate students is extremely small. The number of part-time graduate students is relatively large. The number of part-time graduate students is 54,000 out of 262,000 total graduate students, or approximately 20%. The majority of part-time students are supported in their education by the companies for which they work.

There are three tiers of technical and engineering education (tiers 1 and 2 deliver higher education):

- Tier 1: 4-year universities account for approximately 260 institutions (out of the total of 765); the student age range is 18–22.
- Tier 2: 5-year colleges of technology (64 in total: 55 national, six public prefecture, and three private); the student age range is 15–20. There are approximately 60,000 students.
- Tier 3: technical and engineering high schools; the student age range is 12–15. There are 5500 high schools, 8% of which offer technical and engineering courses.
- The traditional structure of faculties at Japanese universities was based on a chair system. Typically, in engineering, one faculty at most of the major universities had six chairs, and each chair was responsible for teaching and research in a particular discipline. One chair represented a full professor, an associate professor, two research associates, and one technician. This system lasted for about 120 years, but, around 2000, the Ministry of Education started to allow universities to introduce non-chair systems. This was necessary because of dramatic changes in the direction of research, the emergence of a number of new research areas, and the requests of universities to be allowed to establish different interdisciplinary faculties.

- Presently, approximately 260 faculties exist in engineering fields at 166 universities (out of a total of 765): among the 86 national universities, 57 have such faculties; the respective count at the 90 public universities is 18; and at the 589 private universities, 91.
- The number of undergraduate students who are participating in an engineering discipline is 410,000, representing 17% of all such students. The number of students pursuing a master's degree is 65,000 (39.4%). This is by far the largest share among all the disciplines. Social science is second, with 19,000 students. The number of PhD students is 14,000 (18.6% of the total), second after medicine.
- The University Council (1998) identified the roles and functions of universities as follows:
- Pursuing world-class research and education
- Training highly skilled professionals
- Training professionals in a wide range of areas
- Fostering high-quality general education
- Carrying out education and research in specific areas (sports, the arts, and so on)
- Promoting a lifelong learning community
- Contributing to society and service to society

- These roles and functions apply equally to engineering universities and universities with engineering faculties. The council stated that each institution is free to choose one or more of any of these functions. Following the council's report, the Ministry of Education, Culture, Sports, Science and Technology (MEXT) policies became targeted on accelerating a differentiation in the roles and functions of each institution.

- An accreditation-type evaluation system was introduced in 2004. All universities in Japan have to be evaluated once every 7 years. Three evaluation agencies—the National Institution for Academic Degrees and University Evaluation, the Japan University Accreditation Association, and the Japan Institution for Higher Education Evaluation—are responsible for carrying out the evaluations. These agencies have been authorized by MEXT, and their capabilities are monitored by the Central Council for Education. The evaluation covers mainly assessments of teaching quality. The results of the accreditation-type evaluation do not affect funding except in cases of illegal conduct. Each university is free to choose any one of the three evaluation agencies.

- In addition to these evaluations, all national universities must undergo a comprehensive evaluation once every 6 years by an evaluation committee organized by MEXT. The National Institution for Academic Degrees and University Evaluation carries out the evaluation of teaching and research. This scheme was introduced in 2004 following the incorporation of national universities. The evaluation results affect 8–9% of each university's block grant allocation. The impact of this varies from institution to institution. Thus, smaller universities with less alternative funding streams will be affected more than universities with more diversified funding sources.

- As a result of the decline in the size of the youth population, 40–50 private universities are now facing serious economic problems. They may go bankrupt. In such cases, the priority will be to transfer the students in these universities to other universities to ensure minimal disruption in studies (World Bank, 2013, pp. 33–43).

2.3.8.3 Course Structure

Until 1991, there was a strict requirement that the minimum number of credits necessary for a bachelor's degree—124—should include at least 76 credits in specialized subjects, 36 credits in general education subjects, 8 credits in foreign languages, and 4 credits in physical education. As long as a university observes the regulation on the 124 total credits, it is free to structure its curricula. This has led to education reform at Japanese universities, producing a variety of unique curricula structures. Table 2.1 shows the typical curriculum at a small university oriented toward science and engineering.

Table 2.1 Typical curriculum at a small science- and engineering-oriented university in Japan

Required subject areas	Subjects		Credits
Liberal arts	Nonscience	a. Humanities, sociology, literature, law, economics, and so on	Minimum 18, including at least 4 in category a
		b. Interdisciplinary subjects, civilization and culture	
	International communication	Chinese, English, French, German, and Russian	Minimum 14: 10 English +4 in others or 8 English +6 in others
	Health and sports	Health science and physical education	3
Basic science and engineering	Calculus, algebra, physics, chemistry, biology, earth and space science, and descriptive geometry		Minimum 16; the requirement varies by faculty
Specialized subjects			As specified by each faculty
Graduation research project			6–14, varies by faculty
Total			Minimum 124

2.3.9 *Korea*

2.3.9.1 History

Korea developed rapidly from being one of the poorest countries after the Korean War (1950–1953) to now the 11th largest economy in the world underpinned by shipbuilding, semiconductors, PC's, mobile phones, and automobile production. The education system in Korea is strongly based on Confucianism with high value placed on knowledge education and literacy. Seoul University produced Korea's first engineering graduates in 1943.

2.3.9.2 Accreditation

The Accreditation Board for Engineering Education in Korea was chartered in 1999.

An analysis by Jung (2019) of the educational impacts of the Fourth Industrial Revolution in Korea found the greatest challenge to be the industrial structure with the emphasis on reducing production costs to increase profits. The consequence was a reduction in support for universities undertaking collaborative research with industry and the importation of turnkey production packages. Jung (2019) posited:

> In the fourth industrial period, knowledge production should go beyond improving the technology in one field and should rather be a systematic process that includes identifying the problem, creating new designs, prioritizing and making investments, taking risks and scaling up the industries. (Jung, 2019, p. 146)

2.3.9.3 Graduate Attributes

Gleason (2018) foreshadowed a change in the roles of graduate engineers as a result of automation. The so-called soft skills of graduates will be increasingly relevant:

> The new environment also requires the traditional teaching model in higher education to change. Many university graduates were expected to go into middle-skill jobs or managerial positions in the past, but it is estimated that the high proportion of middle-skill jobs will shrink as automation increases. In the previous industrial society, basic disciplinary knowledge and technical education prepared graduates to directly transition into the labour market, but the division of labour based on the disciplinary boundary will be disruptive in the new industrial revolution period. Employers' place more value on soft skills such as creativity, the ability to think outside the box and openness to multiple perspectives because the new labour market is very flexible and transforms quickly. Providing a more cross-disciplinary approach and more diverse educational experiences are emphasised in the changing environment. (Gleason, 2018, p. 147)

2.3.9.4 Course Structure

Typical of the course structure at a Korean university is the civil engineering course at Seoul National University. The first year comprises courses (units) on critical thinking and writing, a foreign language, mathematics, basic science, and computing. These are followed by foundation subjects in statistics with credit points also for general education electives such as college life and leadership. There are discipline-related foundation classes in engineering design and CAD, mechanics of materials, and fluid mechanics. Students then choose five core classes from a list of urban planning, spatial information systems, soil mechanics, traffic engineering, hydrology, environmental engineering, structural analysis, and construction planning and management. Finally, there is a required course on integrated design of civil engineering systems.

2.3.10 Malaysia

2.3.10.1 History

Universiti Malaya is Malaysia's oldest university, founded in 1905 in Singapore as the King Edward VII College of Medicine. It was not until 1954 an engineering department was set up to deliver a 3-year degree with students expected to complete their degrees at overseas universities.

Universiti Teknologi Malaysia (UTM) started as a technical school in 1904. It was elevated to college status in 1955 and then became a technical institute in 1972 and finally a university 1975. As its name implies, UTM's prime focus has been on technology. As a technical college, it delivered diplomas in civil, mechanical, and electrical engineering together with architecture, town and country planning, land surveying, and quantity surveying.

2.3.10.2 Accreditation

Engineering education in Malaysia was reviewed by the Institution of Engineers, Malaysia (EIM), in 2002 following a reduction in some universities' Bachelor of Engineering degree courses from 5 to 3 years post-school year 12. The outcome was a proposed outcome-based model to suit Malaysia's needs for training of professional engineers delivered over a 4- or 5-year period (Johari et al., 2002).

The following five criteria, based on American Accreditation Board for Engineering and Technology (ABET), were considered important in producing graduate engineers after having observed the global engineering models and Malaysia's needs:

- *Scientific strength* provides engineers who are innovative, able to work in research and development activities, and adaptable in different engineering fields.
- *Professional competency* provides engineers who are able to identify, formulate, and solve engineering problems, responsible professionally, and able to use techniques, skills, and modern engineering tools for engineering practice.
- *Multi-skilling* provides engineers who are able to work in different engineering fields and function in multidisciplinary work/teams.
- *Well-respected and potential industry leadership skills* provide engineers who are able to understand the impact of engineering solutions in a global/social context, knowledgeable of contemporary issues, able to communicate effectively, and involved in community or social projects.
- *Moral and ethical soundness* provides engineers who understand their ethical and moral responsibilities.

2.3.10.3 Graduate Attributes

The latest version of the ABET Engineering Criteria contains the same 12 program outcomes (POs), as listed above, developed from the 12 graduate attributes (adopted by all signatories of the Washington Accord) (Sam et al., 2020).

The proposed outcome-based learning represented a significant departure from the traditional content-based approach.

Kok (2020) assessed the effectiveness of the new model; they found the introduction of student-centered learning put additional pressure on both students and teachers. Staff needed to spend additional for preparation and assessment. With the old, didactic, curriculum-based system, the students were used to being "spoon-fed" content and could rely on their memory to pass a unit—the new system tested students' understanding of the subject matter.

2.3.11 New Zealand

2.3.11.1 History

The first higher education institutions to deliver "professional" courses were the University of Otago (medicine in 1875) and the University of Canterbury (engineering in 1887). It was not until 1906 that the Auckland University established a School of Mines which evolved into their School of Engineering with students having to complete their degrees at the University of Canterbury or in the UK. Another university in New Zealand, Massey University, grew out of an agricultural college and delivers engineering degrees in agriculture and food technology.

2.3.11.2 Accreditation

Engineering New Zealand, formerly the Institution of Professional Engineers New Zealand, was formed in 1914 as a collegial body and to oversee standards of professional engineering qualifications.

There are now five other universities in New Zealand offering professional engineering degrees.

2.3.11.3 Course Structure

Some engineering degrees in New Zealand differ from the common 4-year-with 1-year common pattern. Auckland University requires students to take a "General Education" option in the first year. Students can choose from a wide range of non-engineering units: for example, "Introduction to Spanish," "Dancing and Culture," and "Marketing." Canterbury University has only six common units in the first year with civil and mechanical engineering students taking "Chemical Principles and Processes" and "Engineering Mechanics," whereas the electrical and electronics students take "Discrete Mathematics" and "Introduction to Computer Science" instead. Massey University requires students to pass all first-year units before proceeding to second year.

2.3.12 Pakistan

2.3.12.1 History

Major civil engineering works during the time of the British occupation include the railway from Karachi to Kotri built in 1847 and the Lansdowne Bridge across the Indus River in 1889. The first technical college in Pakistan, Mughalpura Technical College in Lahore, dates from 1921.

2.3.12.2 Accreditation

The Pakistan Engineering Council (PEC) was formed in 1976 as the regulatory and accreditation body for professional engineers.

A PEC review of engineering education in 2001 was critical of the ability of the Pakistan higher education system to produce the professional engineers needed of national development. "… of the 24 universities established in Pakistan since the nation was founded only six offer engineering" (Medhi & Rizvi, 2001). In the following 20 years, this number has grown to 135 higher education institutes (PEC, 2024).

The entry point into the engineering profession (a "registered engineer") is a 4-year Bachelor of Engineering or a Bachelor of Science in Engineering degree from a 4-year accredited higher education institution. The accredited courses are highly prescriptive; for example, in the Building and Architecture Bachelor of Engineering program, there is a compulsory unit in computer-aided design which specifies the use of AutoCAD with a lesson on "Create and use layers" (PEC, 2020).

2.3.12.3 Graduate Attributes

The professional engineering degrees in Pakistan have the same program objectives (graduate attributes) as those adopted by the China Engineering Education Accreditation Association (CEEAA) listed above.

2.3.13 Singapore

2.3.13.1 History

The National University of Singapore and the Nanyang Technological University in Singapore are both highly ranked universities internationally. They are ranked 1 and 2, respectively, in the Southeast Asia section of the 2024 QS Asia University Rankings.

The National University of Singapore (NUS) was founded in 1905 with the Faculty of Engineering, now the College of Design and Engineering, formed in 1955. Nine professional engineering specializations are offered: biomedical, chemical, civil, computer, electrical, environmental, industrial and systems, and mechanical. There are also courses in materials science and engineering and engineering science.

The NUS Program Educational Objectives (POEs) are:

1. Depth in fundamental knowledge of core civil engineering disciplines
2. Breadth in integrative skills to apply knowledge across disciplines
3. Appreciation of interactions between engineering, business, and technology in modern society
4. Drive for lifelong learning and continuous self-development
5. Understanding of their role as civil engineers in the sustainable developments of society at the national and global context

2.3.13.2 Accreditation

The Institution of Engineers, Singapore, was established in 1966. Its Engineering Accreditation Board (EAB) is responsible for accrediting all professional engineering degrees awarded by Singapore universities.

2.3.13.3 Graduate Attributes

The Singapore EAB has adopted the same set of graduate attributes (GAs) as China listed above published by the Washington Accord as the basis of student learning outcomes (EAB, 2018).

2.3.13.4 Course Structure

A feature of the NUS engineering degrees is the facility to "build your own degree." Take the Bachelor of Engineering (Civil) as an example; the degree comprises 160 "units" delivered over 4 years. An academic unit, called a "course," is worth 4 units. A typical semester workload is five courses. To build a degree, the student must choose from a wide range of "soft" options such as "Cultures and Connections," "Design Thinking," and "Sustainable Futures" totaling 60 units in the first three semesters. Then follows the compulsory engineering core, engineering calculus, linear algebra, numerical methods, etc. (20 units—one semester), and a "major program" of civil engineering content including fluid mechanics, structural modeling, and foundation systems (40 units—two semesters). Finally, the students choose 40 units from "Unrestricted Selection" to bring the total to 160 units taken over the 4 years.

In contrast, the Nanyang Technological University (est. 1981) offers 49 professional engineering degrees comprising a 4-year Bachelor of Engineering with Honors program but with an additional 20 weeks of industrial placement as a "second major." The BE(Hons) program follows the standard pattern of a common first year of mathematics and physics, as shown in Table 2.2.

The variety of professional engineering degrees at NTU is illustrated by the degree "Bachelor of Engineering in Civil Engineering with a Second Major in Entrepreneurship." Graduates from the 4-year BE(Hons) degree must undertake a 20-week entrepreneurship program outside Singapore or work as a professional intern with a Singaporean start-up or venture capital firm to qualify for this degree.

2.3.14 Sri Lanka

2.3.14.1 History

The University of Ceylon was established in 1942, and in 1953 they awarded the first Bachelor of Science in Engineering degree. The history of engineering education in Sri Lanka is comprehensively recorded by Somasundaraswaran (2009); at that time, there were 15 universities in Sri Lanka including the University of Ruhuna with three universities offering engineering degrees: University of Peradeniya and University of Moratuwa. Engineering courses were delivered in English language, and interestingly the area of specialization was allocated on the basis of student

Table 2.2 First year engineering common engineering curriculum at Nanyang Technological University

Course Code	Course Title	Type	AU	Pre-Requisite
YEAR 1 SEMESTER 1				
MH1810	Mathematics 1	C	3	
PH1011	Physics[a]	C	3	
FE1073	An Introduction to Engineering & Practices	C	1	
EG1001	Engineers in Society	C	2	
CC0001	Inquiry and Communication in the Interdisciplinary World	CC	2	
CC0002	Navigating the Digital World	CC	2	
	Broadening & Deepening Electives 1	BDE	3	
			16	
YEAR 1 SEMESTER 2				
MH1811	Mathematics 2	C	3	MH1810
CV1011	Mechanics of Materials	C	4	
CV1013	Civil Engineering Materials	C	3	
CV1014	Introduction to Computational Thinking	C	3	
CC0003	Ethics & Civics in a Multi-Cultural World	CC	2	
CC0005	Healthy Living and Wellbeing	CC	3	
			18	

[a] Students without 'A' Level Physics will take 'PH1012 Physics A' (4AU)
Note: *C* Core, CC Common Core, *BDE* Broadening and Deepening Elective
Retrieved from: https://www.ntu.edu.sg/docs/librariesprovider117/admissions/curriculum/civil/cee_ce.pdf?sfvrsn=448495fa_7

performance in the first year with equal numbers allocated to electoral and information civil environmental engineering and mechanical civil and environmental engineering.

2.3.14.2 Accreditation

Students completing a full year course graduating with either a Bachelor of Engineering degree or a Bachelor of Science in Engineering degree from a university in Sri Lanka are designated as associate engineers by the Institution of Engineers, Sri Lanka.

Peradeniya University, the largest university in Sri Lanka, now awards BScEng degrees in chemical and process engineering, civil engineering, computer engineering, electrical and electronic engineering, mechanical engineering, and manufacturing and industrial engineering.

The philosophy, aims, and objectives of the course in civil engineering at Peradeniya University are proclaimed by the following mission statement: “The mission of the Faculty is to acquire, promote and disseminate knowledge of engineering sciences and their application to improve the quality of life and, in particular, to equip present and future generations of men and women with skills and attitudes to attain competence as professional engineers, and to interact with industry by providing expertise and specialised facilities."

2.3.14.3 Graduate Attributes

The BScEng graduate specializing in civil engineering is thus expected to:

- Have a sound understanding of theory, laboratory practice, and design and be able to apply this knowledge to the practice of engineering.
- Be capable of using appropriate engineering methods and tools including IT.
- Be capable of carrying out tasks as an individual and as a member of a team in multidisciplinary environments.
- Be able to effectively communicate technical information orally, in the form of drawings and reports, and by audiovisual techniques.
- Be aware of the impact of engineering activity on the social, industrial, and physical environment in Sri Lanka and elsewhere.
- Be able to engage in independent learning and to keep abreast of up-to-date tools and techniques.
- Be able to act professionally and ethically and take responsibility within the limits of competence.
- Be able to integrate the knowledge and understanding of mathematics, science, computer-based methods, design, the economic, social, and environmental context, and engineering practice, to solve engineering problems of a complex nature.
- Be able to manage and supervise other people.
- Possess an appreciation of the role of engineering—especially infrastructure planning and construction engineering—activities in the economic and social development of the country.
- Have the potential to eventually provide leadership and managerial skills for the successful management of engineering and other industrial and commercial enterprises.
- The degree program covers all main engineering topics related to civil engineering and equips the students with essential skills to practice as a professional engineer (retrieved from: https://eng.pdn.ac.lk/civil/undergraduate/aboutundergraduate.php).

2.3.14.4 Course Structure

The course in civil engineering is designed to equip students with a thorough grounding of mathematics, engineering sciences, and technology, with a good understanding of the basic principles of core subjects in civil engineering, embracing theory, analysis, and design.

A good grasp of the fundamentals of mechanical and electrical engineering, computing and software-based applications, communication skills, and complementary studies, including economics, management, health, safety, risk, sustainability, professional ethics, and industrial training in an engineering environment, is ensured.

The course structure is illustrated in Table 2.3.

The Program Educational Objectives (POEs) of the Department of Civil Engineering at Peradeniya University are:

1. Identify, analyze, and solve complex problems based on basic principles of engineering sciences and mathematics.
2. Produce innovative engineering designs and solutions considering functionality, aesthetics, safety, cost-effectiveness, environmental friendliness, and sociocultural aspects.
3. Manage and execute engineering projects of multidisciplinary nature giving due consideration to local community, local industry, and national heritage.
4. Promote themselves as leaders in the international arena in their chosen profession as well as in other interested areas through effective communication, lifelong learning, research, and development activities (retrieved from https://eng.pdn.ac.lk/civil/undergraduate/PEOs.php).

Table 2.3 Course structure of the BScEng in Civil Engineering at Peradeniya University

EXISTING COURSE STRUCTURE : B.Sc. Eng. PROGRAMME IN CIVIL ENGINEERING 2020

SUBJECT AREA		1st SEMESTER	2nd SEMESTER	3rd SEMESTER	4th SEMESTER	5th SEMESTER	6th SEMESTER	7th SEMESTER	8th SEMESTER	CREDIT CORE	CREDIT ELECTIVE
MATHEMATICS	CORE	GP115 : CALCULUS I (3)	GP116 : LINEAR ALGEBRA (3)	EM211 : ORDINARY DIFFERENTIAL EQUATIONS (2) EM213 : PROBABILITY & STATISTICS (2)	EM212 : CALCULUS II (2)	EM315 : NUMERICAL METHODS FOR CIVIL ENGINEERS (2)				14	
	ELECTIVE					[illegible] : OPERATIONS RESEARCH I (2)	[illegible] : OPTIMIZATION (2)	[illegible] : OPERATIONS RESEARCH II (2) [illegible] : OPTIMIZATION (2)			4
STRUCTURES AND MATERIALS	CORE	GP109 : MATERIALS SCIENCE (3) GP110 : ENGINEERING MECHANICS (3)		CE201 : MECHANICS OF MATERIALS I (3)	CE20[illegible] : STRUCTURAL ANALYSIS (3)	CE312 : DESIGN OF STRUCTURES II (3)	CE307 : FINITE ELEMENT METHODS IN SOLID MECHANICS (3) CE306 : DESIGN OF STRUCTURES I (3) CE316 : ADVANCED MECHANICS OF MATERIALS (2)			23	
	ELECTIVE							[illegible] : COMPUTER AIDED STRUCTURAL ANALYSIS & DESIGN (2) [illegible] : CONCRETE TECHNOLOGY (2)	[illegible] : DYNAMICS OF STRUCTURES (2) [illegible] : DESIGN OF STRUCTURES III (2) [illegible] : HIGH RISE BUILDING DESIGN (2)		10
FLUID MECHANICS AND ENVIRONMENTAL ENGINEERING	CORE			CE202 : FLUID MECHANICS I (3)	CE205 : ENGINEERING HYDROLOGY (3)	CE302 : ENVIRONMENTAL ENGINEERING (3) CE305 : HYDRAULICS (3)	CE311 : HYDRAULIC ENGINEERING AND DESIGN (3)			15	
	ELECTIVE						[illegible] : WATER SUPPLY AND WASTE WATER ENGINEERING (2) [illegible] : DISASTER MANAGEMENT (2)	[illegible] : IRRIGATION AND DRAINAGE ENGINEERING (2) [illegible] : HYDRAULIC STRUCTURES (2) [illegible] : INDUSTRIAL POLLUTION CONTROL (2)	[illegible] : ENVIRONMENTAL HEALTH AND SANITATION (2) [illegible] : COASTAL ENGINEERING AND COASTAL ZONE MANAGEMENT (2) [illegible] : INTEGRATED RIVER BASIN MANAGEMENT (2)		16
GEOTECHNICAL ENGINEERING	CORE				CE204 : GEOMECHANICS (3)	CE310 : GEOTECHNICAL ENGINEERING (3)	CE308 : GEOTECHNICAL DESIGN (2)			8	
	ELECTIVE					[illegible] : ADVANCED GEOMECHANICS (2)	CE514 : GROUND IMPROVEMENT AND GEOSYNTHETICS (2) [illegible] : FOUNDATION ENGINEERING (2)	[illegible] : GEOTECHNICAL DESIGN AND CONSTRUCTION (2)	CE 515 : GEOHAZARD MANAGEMENT (2)		10
HIGHWAY AND TRANSPORTATION ENGINEERING	CORE					CE318 : TRANSPORTATION AND HIGHWAY ENGINEERING (3)				3	
	ELECTIVE						[illegible] : TRAFFIC ENGINEERING (2)	[illegible] : HIGHWAY ENGINEERING AND DESIGN (2) [illegible] : TRANSPORTATION PLANNING (2)	[illegible] : TRAFFIC MANAGEMENT (2)		8
CONSTRUCTION AND MANAGEMENT	CORE				CE209 : BUILDING CONSTRUCTION (3) MA201 : ENGINEERING MANAGEMENT (3)			CE403 : CONSTRUCTION MANAGEMENT (3)		9	
	ELECTIVE								[illegible] : CONSTRUCT EQUIPMENT AND MATERIAL MANAGEMENT (2) [illegible] : SUSTAINABLE DESIGN AND CONSTRUCTION (2) [illegible] : CONSTRUCTION PLANNING (2)		6
SURVEYING	CORE & ELECTIVES		GP118 : BASIC ELECTRICAL & ELECTRONIC ENGINEERING (3)	CE213 : ENGINEERING SURVEYING (3)			CE317 : CIVIL ENGINEERING FIELD WORK (3) (OUTSIDE SEMESTER)	[illegible] : GIS AND RS FOR CIVIL ENGINEERS (2)		6	2
OTHERS	CORE	GP112 : ENGINEERING MEASUREMENTS (3) GP114 : ENGINEERING DRAWINGS (3) GP101 : ENGLISH I (2)	GP111 : ELEMENTARY THERMODYNAMICS (3) GP113 : FUNDAMENTAL OF MANUFACTURE (3) GP106 : COMPUTING (3) GP102 : ENGLISH II (3)	EE280 : INTRODUCTION TO ELECTRICAL ENGINEERING (3) ME202 : MECHANICAL ENGINEERING FOR CIVIL ENGINEERS (3)	CE219 : CIVIL ENGINEERING LABORATORY I (1)	CE319 : CIVIL ENGINEERING LABORATORY II (1)	CE326 : CIVIL ENGINEERING LABORATORY III (1)	CE405 : CIVIL ENGINEERING PROJECT I (3)	CE406 : CIVIL ENGINEERING PROJECT II (3) CE 402 : MULTI DISCIPLINARY DESIGN PROJECT (3)	42	
	GENERAL ELECTIVE						GENERAL ELECTIVE - 1 (2)	GENERAL ELECTIVE - 2 (2) GENERAL ELECTIVE - 3 (2)	GENERAL ELECTIVE - 4 (2)	RATIO OF TOTAL ELECTIVE CREDITS OFFER AND CREDITS COUNTED 16/56 = 0.29	
CREDITS		18 (CORE)	18 (CORE)	19 (CORE)	18 (CORE)	18 (CORE) + 4 (TE) = 22	17 (CORE) + 6 (TE) + 2 (GE) = 25 3 (CORE OUTSIDE SEMESTER)	6 (CORE) + 6 (TE) + 4 (GE) = 16	6 (CORE) + 4 (TE) + 2 (GE) = 12	120 (CORE) + 16 (ELECTIVES) + 8 (GE) + 6 (INDUSTRIAL TRAINING) = 150	

DATE OF LAST EDIT [illegible]

2.3.15 *Taiwan*

2.3.15.1 History

The oldest university in Taiwan, the University of Taipei, was established in 1895 at the start of the Japanese occupation which lasted until 1945. At the end of World War II, the Japanese system of higher education in Taiwan was replaced by that of China which had followed the American model. Engineers hold a highly respected position in Taiwanese society, and for that reason there are many engineering schools training engineers from technician to PhD level. The most prestigious university is the National Taiwan University (NTU) which has a Times Higher Education World Ranking (2024) of 152; NTU is ranked number one in Taiwan.

2.3.15.2 Accreditation

The Institute of Engineering Education, Taiwan (IEET), established in 2003 accredits 4-year engineering degree programs for recognition with the Washington Accord. The accreditation is outcomes based with a 6-year cycle.

2.3.15.3 Graduate Attributes

The IEET has adopted a different set graduate attributes (IEET, 2015). The program must demonstrate that students have attained the following outcomes by graduation:

- Ability to apply knowledge of mathematics, science, and engineering
- Ability to design and conduct experiments, as well as to analyze and interpret data
- Ability to apply techniques, skills, and modern tools necessary for engineering practice
- Ability to design an engineering system, component, or process
- Ability to manage project (including budgeting), communicate effectively, work in multidisciplinary environment, and function on teams
- Ability to identify, formulate, research literature, and analyze complex engineering problems reaching substantial conclusions
- Knowledge of contemporary issues; an understanding of the impact of engineering solutions in an environmental, societal, and global context; and the ability and habit to engage in lifelong learning
- Ability to apply ethical principles and commit to professional ethics and responsibilities and norms of engineering practice and a sense of respect for diversity

2.3.15.4 Course Structure

For the design and contents of the curriculum to be consistent with the Program Educational Objectives (PEOs) it must demonstrate through transcript analysis that the coursework of each graduate includes the following three major components: mathematics and basic sciences, technical and professional engineering component, and general education. Specifically:

- Mathematics and basic sciences must account for at least 9 credits and total to at least one fourth of the credits required for graduation.
- Technical and professional engineering component must account for at least three eighths of the credits required for graduation including the capstone design project.
- General education component must complement the technical contents of the discipline and be consistent with the PEOs.
- Design and implementation of the curriculum must correlate with the development of the industry and prepare students to culminate the learned knowledge and skills in engineering practice (IEET, 2015).

The NTU has Departments of Civil, Mechanical, Chemical, Engineering Science and Ocean Engineering, Materials Science and Engineering, and Biomedical Engineering delivering 4-year Bachelor of Engineering degrees.

Within the Department of Civil Engineering at NTU, students can specialize in civil engineering, railway engineering, architectural engineering, environmental engineering, natural disaster prevention, and wood construction. The department has the following educational goals:

- Develop basic professional knowledge and skills.
- Develop basic abilities in practical execution, communication, and coordination.
- Develop basic abilities to engage in research.
- Cultivate humanistic qualities and the ability to serve society.

Students can choose between instructions in the Chinese or English languages. A feature of the NTU degree is designating subjects as "cornerstone" (second year), "keystone" (third year), and "capstone" (final year). Otherwise, the BE curriculum in Taiwan is similar to most of the other universities in the Asia-Pacific region with a first year common to all engineering specializations.

References

CEEAA. (2022). *Engineering Education Accreditation Criteria*. Retrieved from: https://www.ceeaa.org.cn/jyzyrzxh/standards/criteria/631908/index.html

Chinna Balu, C. (2014). Engineering education in India an overview. *International Journal & Magazine of Engineering Technology, Management and Research, 1*(8), 7–9.

EAB. (2018). *Engineering accreditation board manual*. Institution of Engineers. Retrieved from: https://www.ies.org.sg/Tenant/C0000005/PDF%20File/Accreditation/EAB%20Version%20Aug%202018.pdf

Gleason, N. W. (2018). Singapore's higher education systems in the ears of the fourth industrial revolution: Preparing lifelong learners. In N. W. Gleason (Ed.), *Higher education in the era of the fourth industrial revolution* (pp. 145–169). Palgrave Macmillan.

Gregory, M. S. (1971). *History and development of engineering*. Longman.

HKIE. (2013). *Professional Accreditation Handbook (Engineering Degrees) (2013 version)*. Retrieved from: https://www.hkie.org.hk/en/quali/criteria/upload/page/55/self/59ba2b4d548ec.pdf

IABEE. (2024). *Common criteria for engineering programs*, Retrieved from: https://iabee.or.id/en/accrediation/accreditation-criteria/common-criteria/

IEA. (2021). *International engineering alliance graduate attributes & professional competencies*. Retrieved from: https://www.ieagreements.org/assets/Uploads/IEA-Graduate-Attributes-and-Professional-Competencies-2021.1-Sept-2021.pdf

IEET. (2015). *Accreditation criteria for accrediting engineering programs*. Retrieved from: https://www.ieet.org.tw/FileSystem/htmlEditor/e77da57c-7409-4c76-a5c2-a60ef19e09c5/0101_Accreditation%20Criteria%20for%20Accrediting%20Engineering%20Programs.pdf

Johari, M. M. N. M., Abdullah, A. A., Osman, M. R., Sapuan, M. S., Mariun, N., Jaafar, M. S., & Rosnah, M. Y. (2002). A new engineering education model for Malaysia. *International Journal of Engineering Education, 18*(1), 8–16.

Jung, J. (2019). The fourth industrial revolution, knowledge production and higher education in South Korea. *Journal of Higher Education Policy and Management, 42*(2), 134–156. https://doi.org/10.1080/1360080X.2019.1660047

Kok, C. W. (2020). A Malaysian outcome-based engineering education model: The implementation and challenges in future. *INTI Journal, 2020*(20).

McNeill, N. Li, S. & Wang, J. (2010). *Growing Pains: Chinese Engineering Education During The Late Qing Dynasty*. School of Engineering Education Graduate Student Series. Paper 18. http://docs.lib.purdue.edu/enegs/18

Medhi, M. R., & Rizvi, S. A. I. (2001). A recent review of engineering education in Pakistan. *Journal of Engineering Education*, 207–211.

PEC. (2020). *Pakistan engineering council—curriculum for building and architectural engineering*. Bachelor of Engineering Degree. Retrieved from: https://www.pec.org.pk/wp-content/uploads/2021/03/Building-Architectural-Engineering.pdf

PEC. (2024). *Pakistan Engineering Council - First Schedule (programs accredited under level-II, substantially equivalent to Washington Accord)*. Retrieved from: https://www.pec.org.pk/accredition/programs-under-level-2/

Sam, M. K., Cai, S. K., & Wen, K. C.(2020). Malaysian Outcome-Based Engineering Education Model: The Implementation and Challenge in Future. In *International Conference on Innovation and Technopreneurship*.

Somasundaraswaran, K. (2009). An overview of engineering education in Sri Lanka. In *20th Australasian Association for Education Conference, Adelaide*.

SWJTU. (n.d.). *Establishment and a Rise to Fame (1896–1949)*. Retrieved from: http://civil.ix2.com/vatuu/EnglishView?viewType=About&jsonpName=EAA13A12034EC760F218AA17A40484A6

University Council. (1998). *A Vision of Universities in the 21st Century and Reform Measures*. Report (October 26), Tokyo, Ministry of Education, Culture, Sports, Science, and Technology.

World Bank. (2013). *International Comparative Study : Engineering Education in India*. South Asia Human Development Sector report no. 57;. © Washington, DC. http://hdl.handle.net/10986/16562 License: CC BY 3.0 IGO.

Wu, K. B. (1992). *Investment in science and technology during the time of political and economic change. Education and employment division population and human resource is department*. The World Bank. PHREE background paper series document number PHREE/92/70 130p.

Chapter 3
Training Professional Engineers in Australia

Abstract This chapter focusses on the training of professional engineers in Australia drawing on the experience of the author who enrolled as a student in the Faculty of Engineering at the University of Tasmania in 1957 at a time when slide-rules and Imperial units were in use and there were 3 years of the course common to the Civil, Mechanical and Electrical specializations. The Faculty of Engineering at the University of Tasmania is taken as a case study of the changes in curricula and learning and teaching styles in engineering education which followed. With the advent of micro-electronics and computers in the early 1960's the course needed to change to accommodate these new fields. The 3 years' common course was reduced to 2 years and is now only 1 year common to all engineering undergraduates. How this transpired and the internal politics of curriculum change is covered in detail in Chap. 4. How the curricula, the learning styles and the laboratories and workshop facilities combine to provide the competencies required of an engineering graduate are explored. The chapter includes discussion on assessment methods, marking rubrics and the concerns of academic integrity. The alignment of intended learning outcomes, learning tasks and assessments is explained. The chapter concludes with comment on blended learning and the nexus between teaching and research.

3.1 Summary

This chapter explores the way professional engineers are trained in Australia with a case study of the University of Tasmania (UTAS) where the author of this book, Dr. Peter Doe, completed Bachelor of Science and Bachelor of Engineering with Honours degrees in 1963. Dr. Doe worked as an engineer with the Hydro-Electric Commission in Tasmania before returning to his *alma mater* to complete a PhD part time while lecturing in mechanical engineering. He spent a year at the University of Reading's National School of Food Technology on study leave and was Head of

P. E. Doe, *Prospects for Professional Engineering Education in the Asia-Pacific Region*, Education in the Asia-Pacific Region: Issues, Concerns and Prospects 74,
https://doi.org/10.1007/978-981-96-3197-1_3

Engineering at the Australian College of Kuwait for 2 years after retiring as Associate Professor in Mechanical Engineering at UTAS. His time as an engineer and academic started when computers replaced slide rules; learning and teaching changed from the traditional Confucius-style didactic delivery to student centred- and on-line learning. The common thread throughout has been the reliance of laboratory work for students to develop ‘hands on’ skills, although this is changing with the emergence of computer simulation.

The first Professor of Civil and Mechanical Engineering in the University of Tasmania, Archibald “Arch” Oliver was appointed in 1955. Professor Oliver had a vision for engineering education that lecture material should be built on a foundation of practical skills and knowledge. As a result, the new engineering building on the Sandy Bay campus, erected in 1956, has generous laboratory and workshop spaces. Part of Professor Oliver’s manifesto for the training of professional engineers is published below.

The Australasian Association for Engineering Education held its first conference in 1989. Since that time academics, support staff, postgraduate students, librarians, professional engineers and employers from the SE Asian region and beyond have met to share their experience and knowledge. Their aim is “to improve the quality, relevance and performance of engineering education in Australasia through various events, conferences, awards and professional development opportunities”.

The style of learning and teaching at universities in Australasia changed dramatically in the second decade of the twenty-first century. First there was the change to student-centred teaching. Constructive Alignment followed (Biggs, 2014) with delivered material aligned to Intended Learning Outcomes (Biggs & Tang, 2011), Assessment Tasks, and Engineers Australia Stage 1 Competencies for Professional Engineers (Engineers Australia, n.d.).

As a result of the COVID-19 pandemic there was a rapid uptake of on-line delivery with lecturing staff and ICT technicians scrambling to facilitate lectures on Zoom. In the later months of the pandemic there was a reluctance to return to face-to-face delivery. This was partly from students becoming comfortable with off-campus study and the opportunities afforded for part-time work, but also because university administrators discovered efficiencies in space utilization (not needing large lecture theatres) and staff productivity (recycling recorded lectures).

The outcome, post COVID-19, at the University of Tasmania mandated blended-learning with content material posted on-line and face-to-face restricted to tutorials, workshops and laboratory sessions. The old-style 50-minute presentations to a lecture theatre full of students have disappeared, except for lectures from distinguished persons, or Vice-Chancellors.

3.2 Engineering Fundamentals

The training of professional engineers evolved from the 'Master-apprentice' model to formal class- and institution-based instruction building on the fundamentals of mathematics and science leading to a knowledge and understanding of codes of practice and practical skills.

Under the 'Master-apprentice' model a young man or woman would work alongside a Master as a blacksmith, carpenter, candlemaker, watchmaker, silversmith and other trades for 8–15 years to acquire the necessary skills to become a Master and, in turn, train an apprentice. When the Industrial revolution began in the middle of the nineteenth century there was a change in the model. More technical, as distinct from practical, skills were needed to understand and advance the new technology.

The practice of alchemy—the attempt to make gold from common metals was superseded by the emerging science of chemistry. The periodic table of the elements was formulated by Mendeleev in 1869. Two centuries earlier, understanding of thermodynamics came from the work of Robert Boyle, Jacques Charles, and Joseph Louis-Gay-Lussac after whom the 'Perfect Gas' laws were named. Every engineering student who has studied thermodynamics will know of the name Sadi Carnot epitomised in the 'Carnot Cycle' for heat engines.

Likewise, the science of mathematics blossomed in the seventeenth centuries with the invention of logarithms (Napier) and Calculus (Newton and Leibnitz).

Understanding how stresses in bridges and tunnels relate to applied loads became important with the advent of railways. The time-honoured technique of making bridges in stone or brick heavier to support higher loads no longer applied to the newly adopted materials wrought iron, cast iron and steel. Bridge designers needed to understand how to calculate the size of the structural members and how to connect them using rivets or welds. They also need to know about geomechanics—the strongest dam or bridge will fail if its foundations are flawed.

The design of the mechanical components of machines; gears, linkages, cams, springs did not change much from ancient times until the beginning of the twentieth century until stronger and lighter materials were developed for aircraft. Engineers had to know how to use aluminium, titanium, carbon-fibre and graphene.

Aerodynamics and fluid mechanics became important with the design of aircraft, canals, sewerage, and hydro-electric schemes. Engineers needed to understand boundary layers, hydraulic jumps and aircraft stall. Space travel and supersonic speeds are now commonplace.

It was not until late in the nineteenth century that electrical engineering emerged as a profession. Cities around the world replaced gas-lamps with electrical lights. Telegraph communication was superseded by radio. Thermionic valves were developed only to make way for semiconductors in the middle of the twentieth century. Radar and television followed but the game-changing development was the computer which replaced slide rules, log tables and mechanical calculators for engineers in the mid 1960s.

With computers came the new disciplines of Computer Science, and Information and Communications Technology (ICT).

3.3 Stage 1 Competencies

The Engineers Australia Stage 1 Competency Standard for Professional Engineer (Engineers Australia, n.d.) specifies the knowledge and skills base for graduate engineers. Without specifying areas of study, graduates are expected to have a 'Comprehensive, theory-based understanding of the underpinning natural and physical sciences and the engineering fundamentals applicable to the engineering discipline'. This element of competency requires, as an indicator of attainment, that the graduate 'Engages with the engineering discipline at a phenomenological level, applying sciences and engineering fundamentals to systematic investigation, interpretation, analysis and innovative solution of complex problems and broader aspects of engineering practice.'

There is also required a 'Conceptual understanding of the mathematics, numerical analysis, statistics, and computer and information sciences which underpin the engineering discipline'. For this competency the indicator of attainment is: 'Develops and fluently applies relevant investigation analysis, interpretation, assessment, characterisation, prediction, evaluation, modelling, decision making, measurement, evaluation, knowledge management and communication tools and techniques pertinent to the engineering discipline'.

The challenge, therefore, in designing a curriculum for an engineering degree, is to decide which of the subject areas should be taught. Must all prospective engineers be exposed to the fundamentals of chemistry, mathematics, stress analysis, mechanics, thermodynamics, fluid mechanics, materials, electrical power, electronics, communications and ICT? In the pre-computer age students would learn facts to be regurgitated at exam time. Now much of this detail is readily available from computer search engines at the click of a button. A solution, since the middle of the last century, was to introduce specialisations. An electrical engineer is not taught soil mechanics. A structural engineer needs not to know Maxwell's Equations for electrical and magnetic fields.

Attaining the EA Stage 1 competencies at a university or technical college is only the first step in the training of an engineer. There is the expectation for new graduates (Graduate member of Engineers Australia for students who have satisfied the EA Stage 1 Competencies of a Professional Engineer) to progress to full membership of Engineers Australia (Chartered Engineers) after several years working as an engineer to acquire the EA Stage 2 competencies. Throughout their working life professional engineers engage in further study to keep abreast of current developments and practice. Continuous Professional Development (CPD) is a requirement for on-going registration with accrediting bodies.

3.4 'Soft Skills'

Most engineers, after graduation and becoming chartered engineers, have a management role. This role may be as small as overseeing the foreman of a team of construction workers or as the CEO of a major industry. In many cases the work of a professional engineer is only marginally related to the technical matter they studied in their degree. Many instances of engineering graduates progressing to managerial roles are presented in Chap. 5.

Employers of engineering graduates expect them to be proficient communicators. Hence the 'job interview' prior to appointment. There is an element in the EA Stage 1 Competencies which is directly related to the so-called 'soft skills'. Graduates are expected to demonstrate 'Effective oral and written communication in professional and lay domains'. The indicators of attainment are:

a) **Is proficient** in listening, speaking, reading and writing English, including:—comprehending critically and fairly the viewpoints of others;—expressing information effectively and succinctly, issuing instruction, engaging in discussion, presenting arguments and justification, debating and negotiating—to technical and non-technical audiences and using textual, diagrammatic, pictorial and graphical media best suited to the context;—representing an engineering position, or the engineering profession at large to the broader community;—appreciating the impact of body language, personal behaviour and other non-verbal communication processes, as well as the fundamentals of human social behaviour and their cross-cultural differences.
b) **Prepares** high quality engineering documents such as progress and project reports, reports of investigations and feasibility studies, proposals, specifications, design records, drawings, technical descriptions and presentations pertinent to the engineering discipline. (Engineers Australia, n.d.)

The challenge, in formulating a curriculum for an engineering degree, is to include content whereby students can acquire these 'soft skills'. Requiring students to write reports of laboratory experiments is part of the solution. Another approach is to require students to present their designs, for example, in the form of a presentation to their peers and staff. Such presentations often are part of the assessment in a unit. In the following section on Curriculum Design it can be seen how several units contribute the 'Oral and Written Communication' competency element.

In 1998 I was President of the Tasmania Division of the Institution of Engineers, Australia. When my term as President expired, I gave a valedictory speech on the importance of engineers being skilled communicators. Here is part of this speech reproduced here to illustrate the importance to an engineer (and in my case an academic) of communication skill:

> As a Head of Department, I have no less than 8 ways of communicating with my staff:
>
> - Telephone
> - E-mail
> - Memorandum

- Facsimile
- Formal letter
- Notice on the notice board
- Verbal message relayed via the secretary (when you see John, tell him that I want to see him) and,
- Getting off my backside and going to John's office.

The importance of effective and socially relevant communication cannot be overemphasized. We have all heard the classic example of the wartime message "send reinforcements we're going to advance" ending up as "send three-and-fourpence we're going to a dance" and can remember the game where a whispered message is passed around a circle.

But effective communication is more than getting an accurate transmission of text.

For this talk, for example, I have a choice of several methods of delivery:

I could (heaven forbid) have passed out the text of my talk and be standing here reading it while you do the same. The noise of you changing pages is deafening.

Or I could be reading dot points from prepared slides; which are often typed in 12 point so that not even the author can read them from the screen.

I could have memorized the entire script, or parts of it, and recite it as if I was acting in someone else's play.

Or, having done the preparation, and trusting on my lecturing experience, I could have come along with a brief outline, and more or less 'winged' it, which is what I am doing now.

The choice of method is influenced by the depth of the lecturer's knowledge, experience, audience expectation (those attending an advertised 'public lecture' would expect something different to a chat session which would be appropriate to a university tutorial). There is also the situation where a talk can be taken literally as a definitive statement—for example a politician's answer to a question on notice. These are invariably read from a prepared text so that there can be no slips or confused interpretation.

Getting back to my message to staff... Let's say that I have to break the news that, because of budgetary constraints, we will all be required to take our annual leave from January 4 to January 18. This should best be done at a staff meeting, where we could all discuss the implications and decide if any action in response was appropriate or worthwhile.

But what if the bad news affected only one or two staff, the secretaries, for example. The message would still be the same, but it would be much better delivered to those affected individually and in private.

Even fairly straight forward messages, such as an announcement that Professor John Wellregarded, the Professor of Civil Engineering from the University of Southern Sydney will be visiting the School next Wednesday afternoon, can be delivered in different ways with different effects. A notice on the notice board raises a level of awareness, but little more. A telephone call to staff, depending on what they are doing at the time, might have a similar lack of impact. But an E-mail giving a brief outline of Professor Wellregarded areas of expertise with a list of individuals 'volunteered' to show him around at different times is likely to achieve the desired result.

Even in standard English there are subtle shades of meaning that can be misunderstood.

What would a non-English speakers make of the instruction:

'You can make these footings larger'.

Could they distinguish between this and.

'You **could** make these footings larger'**,** or.

'You **should** make these footings larger'.

When what you really mean is 'You **must** make these footings larger'.

or more simply 'Make these footings larger'.

Not all communication is verbal. Body language plays an important part in communication. As anyone who has been to certain parts of the sub-continent will know it is quite disturbing for someone to be agreeing with you and shaking their head from side to side.

Communication is, or should be, a two-way process. Being a good listener is just as important a skill as being a good speaker. The whole area of negotiation relies on the adoption of a suitable communication style. Passive communication, submissive or 'nice', putting up with things in order not to rock the boat, will result in your needs unlikely to be met. On the other hand, an aggressive communication approach may work in the short term, but it generates anger and hostility. Assertive communication, being able to say what you want, need, think and feel, but not at the expense of others; saying the right thing, in the right way, at the right time—is more likely to result in a good outcome. The technique for assertive negotiation is firstly to show you listen and understand; say what you think or feel; then say what you want to happen: Think before you speak, and use words like 'however' rather than 'but'.

There are also some tricks which public speakers use to reinforce their message: You often hear the 'rule of three' used. For example: 'Friends, romans, countrymen, lend me your ears' or another technique, repetition of phrases (anaphora) "We shall fight on the beaches, we shall fight on the landing-grounds, we shall fight in the fields and in the streets, we shall fight in the hills". Good orators use these tricks to good effect.

Why should engineers, managers and university teachers be good communicators? The answer can be seen from a contemplation of the effects of bad communication. Anything from having a building fall over because the footings are not large enough, to Professor Wellregarded turning up on Wednesday afternoon when all the staff have gone off to play golf.I challenge you to listen critically how you communicate and to decide whether, in certain circumstances, you could be misunderstood. (Doe, 1998).

3.5 Curriculum Design

Typically, the engineering degree curriculum is designed to deliver technical skills based on a strong foundation of mathematics and physics. Most engineering degrees require, as prerequisites for admission, students be competent with algebra, trigonometry and graphing. In Australia the advanced mathematics syllabus at high school also includes the exponential and logarithm functions, some statistical analysis and an introduction to differential and integral calculus although calculus is usually included in a first-year engineering degree mathematics unit. Advanced physics taught at high school includes thermal, nuclear and electrical physics, linear and wave motion, Newton's Laws, gravity and electromagnetism and an introduction to special relativity.

Not all students enrolling in an engineering degree course have studied mathematics and physics at the highest level in high school. Also, there are (mainly mature age) students admitted on the basis of their experience working in an engineering industry or have a TAFE qualification. The first-year curriculum should fill any gaps in students' understanding of mathematics and physics and provide a foundation for the engineering specialisations studied in the following years.

To give an example, the first year Bachelor of Engineering degree course at the University of Auckland in New Zealand is a common year for all engineering specialisations. It comprises:

- Engineering Mechanics (15 points)
- Energy and Society (15 points)
- Materials Science (15 points)
- Electrical and Digital systems (15 points)
- Engineering Computation and Software Development (15 points)
- Principles of Engineering Design (15 points)
- English Language Competency (no credit)
- Academic Integrity Course (no credit)
- General Education (15 points)

By way of contrast, the first year at Sydney University has a flexible approach in which, as well as foundation units in engineering computing and mathematics, students are introduced to each of the eight streams in their first semester: aeronautical-, biomedical-, chemical and biomolecular-, civil-, electrical-, mechanical-, mechatronics- and software-engineering. They then choose a specialisation.

Another example is the Bachelor's Degree programme at the National University of Singapore. Students choose their specialisation on entry: biomedical-, chemical-, computer-, electrical- engineering science-, environmental, industrial design, industrial and systems engineering, materials science, or mechanical-engineering. They then choose from a wide range of common curriculum, major, second major, minor and specialisation electives.

3.6 Curriculum Mapping

Carew et al. (2013) introduced the concept of curriculum mapping of engineering degree courses.

Mapping of Intended Learning Outcomes (ILOs) to EA Stage 1 competencies was used to support the University of Tasmania's 2013 application to Engineers Australia for accreditation of Bachelor of Engineering degrees.

Table 3.1 is the mapping for all units in the Mechanical Engineering specialisation at the University of Tasmania in 2013.

Every unit in each of the specializations of the School of Engineering Bachelor of Engineering degree contributes to development of one or more of the EA Stage 1 Competencies.

Unit coordinators rated their assessment against the EA Stage 1 Competency standards on a scale of 1 to 3 where:

1. the element of competency had been **indirectly** assessed; the subject matter, activities, teaching program and learning outcomes were designed to develop the competence,
2. the element of competency had been **partially** assessed, and,
3. the element of competency had been **directly** assessed in this unit.

In no instance was there a claim for competency that was assessed but not taught.

Table 3.1 EA Stage 1 Competencies gained from each unit in the UTAS Bachelor of Engineering (Mechanical) degree (SoE, 2013)

	Unit code	Engineers Australia Stage 1 Competencies															
		1.1	1.2	1.3	1.4	1.5	1.6	2.1	2.2	2.3	2.4	3.1	3.2	3.3	3.4	3.5	3.6
Year 1																	
Calculus and Appl. 1A	KMA152	3	3														
CAD and Commun.	KNE111			3	2								2	2		1	
Prog. &Prob. Solving	KXT101			3					3								
Engineering Statics	KNE113	3							2								
Calculus and Appl. 1B	KMA154	3	3														
Eng.Circuit Theory	KNE122	3	3														2
Engineering Dynamics	KNE151	3							2								
Chemistry of Materials	KRA170	3	3														
Year 2																	
Eng. Mathematics	KME271	3	3														
Civil Engineering 1	KNE216	3	3						2								1
Electrical Eng. 1	KNE223	3							2								
Mech. Engineering 1	KNE251			3					3								
Materials & Manufac.	KNE210				3				3								
Eng. Des. & Proj. Man.	KNE211				3			3		2	2			2		1	
Thermal and Fluid Eng.	KNE213							3	3				3				
Reliability Eng	KNE240		3			1	3		3					1			
Year 3																	
Eng. Num. Methods	KME300	3		3					3			3			2		
Thermal Engineering 1	KNE354							3	3								
Mech & Mechatronic	KNE356						2	3		3				3			
Dynamic & Mech Syst.	KNE357							3	3	3							
Structural Mechanics	KNE312			3				3	3								1
Control systems Eng	KNE336	3						3	2						1		
Prof. Eng. Practice	KNE346				2	2					2	1					1
Fluid Mechanics 1	KNE351			3				3	3				1				
Industrial Experience	KNE380										1		2			2	
Year 4																	
Stress Analysis	KNE412			3				3	3								
Fluid Mechanics 2.	KNE451			3					3		3		2				
Environmental Eng.	KNE470								3	3	3	3			3		
Engineering Project A	KNE489			3	3	3	3			3	3		3		3	3	
Thermal Engineering 2	KNE454							3	3				3				
Mech Design & Manuf.	KNE457					3			3		3		2				
Adv Dynamic & Vibr.	KNE462							3	3			3					
Engineering Project B	KNE490			3	3	3	3			3	3		3		3	3	
Number of units claiming competency delivered		11	7	10	6	5	4	11	21	6	8	4	9	4	5	5	4

The assessment rating for each competency is on a scale of 1 to 3 where:

3 - Indicates that this element of competency has been directly assessed in this unit

2 - Indicates that this element of competency has been partially assessed

1 - Indicates that this element of competency has been indirectly assessed. The subject matter, activities, teaching program and learning outcomes are designed to develop this competence.

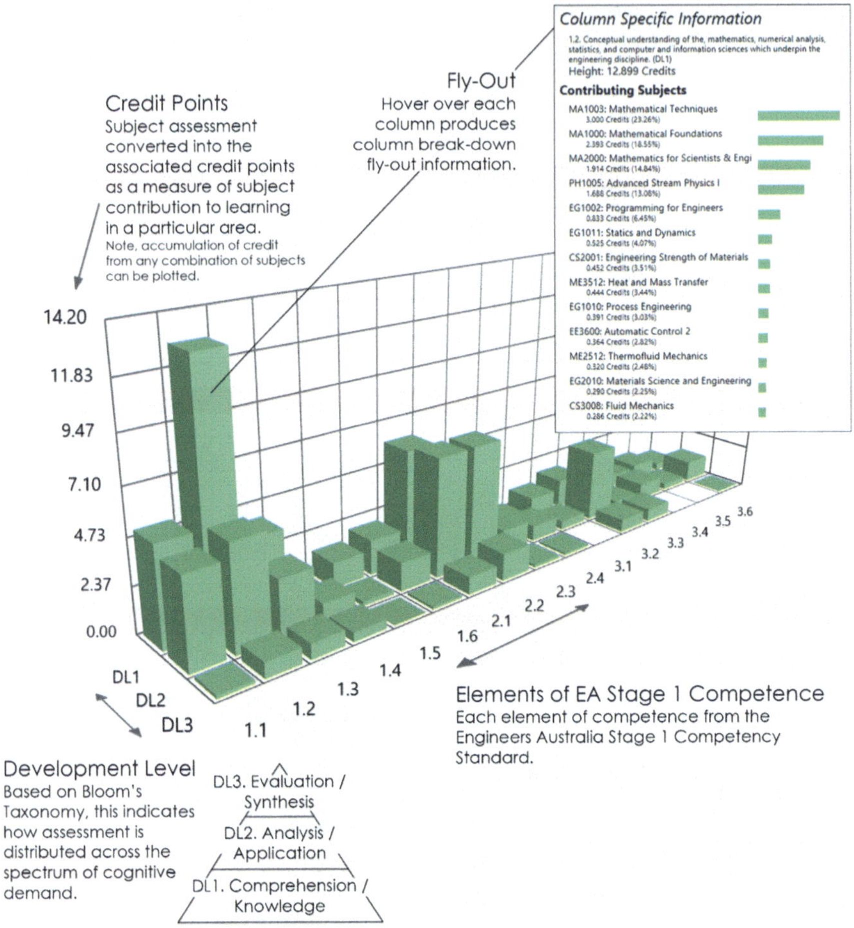

Fig. 3.1 Mapping result for the Mechanical Engineering Program at James Cook University (Holmes et al., 2018)

Most of the competencies claimed by a particular unit were assessed directly by laboratory reports and formal (supervised) examination. Some were partially assessed; for example, in the unit KNE211 *Engineering Design and Project Management* EA Competency *3.5 Orderly management of self, and professional conduct* was partially assessed. Students had multiple tasks to prepare and report on a strict deadline with penalties for late submission. This necessitated prioritization given the requirements of other units.

Some competencies were claimed indirectly, for example in the unit KNE240 *Reliability Engineering*, Competency *3.3 Creativity, innovative and pro-active demeanor* was indirectly assessed: The subject matter and in-class activities were designed to develop this competence.

Holmes et al. (2018) extended this methodology to map attributes of a mechanical engineering degree at another Australian university to the EA Stage 1 competencies. They claim the method is 'equally applicable to a wide range of professional programs across Australia, Europe and the rest of the world'. The dimensions of Credit Points and a Development Level (DL) which is based Bloom's Taxonomy are mapped to the EA Stage 1 Competencies in a three-dimensional bar plot (Fig. 3.1).

3.7 Constructive Alignment

In 2016 the Constructive Alignment framework of Biggs and Tang (2011) was adopted for all units and courses at the University of Tasmania.

> "Constructive alignment is a design for teaching in which what it is intended students should learn and how they should express their learning is clearly stated before teaching takes place. Teaching is then designed to engage students in learning activities that optimise their chances of achieving those outcomes, and assessment tasks are designed to enable clear judgments as to how well those outcomes have been attained" (Biggs, 2014, pp. 5–6).

The application of constructive alignment to a subject requires:

- Identification of the intended learning outcomes (ILOs)
- Designing assessment tasks to measure attainment of the learning outcomes,
- Planning learning activities to enable students to develop the skills, knowledge and understandings described in the ILOs and measured by assessment, and
- Choosing the content (topics/examples/resources/materials) required to support the learning activities.

3.8 Problem Based Learning

More recently there has been a shift from traditional content–based subjects to experiential or problem-based learning. Students are set a design, or a design-and-build task, which requires for its solution an understanding of the fundamental principles. The classic example is for first year students to design (Fig. 3.2), analyse (Fig. 3.3) build and test (Fig. 3.4) a wooden bridge.

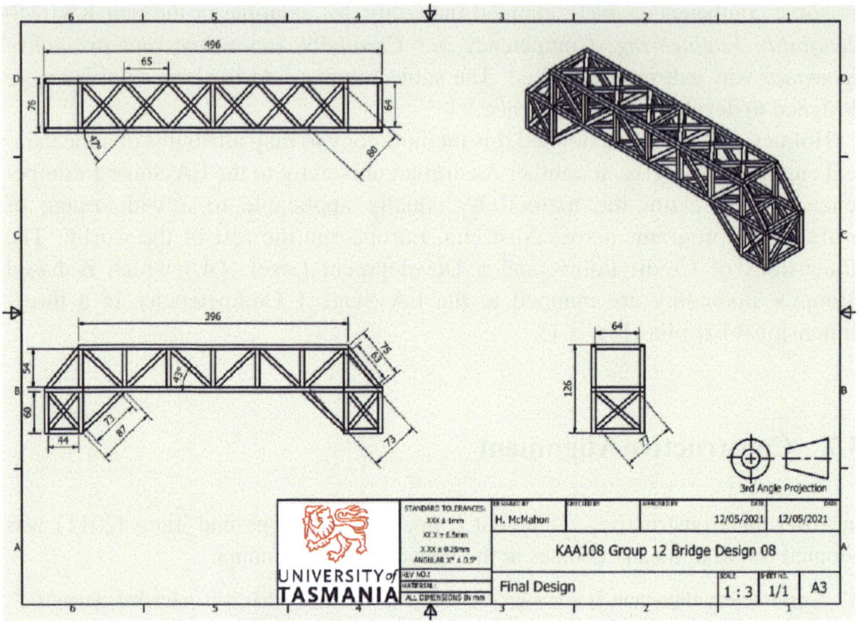

Fig. 3.2 Design of a model wooden bridge

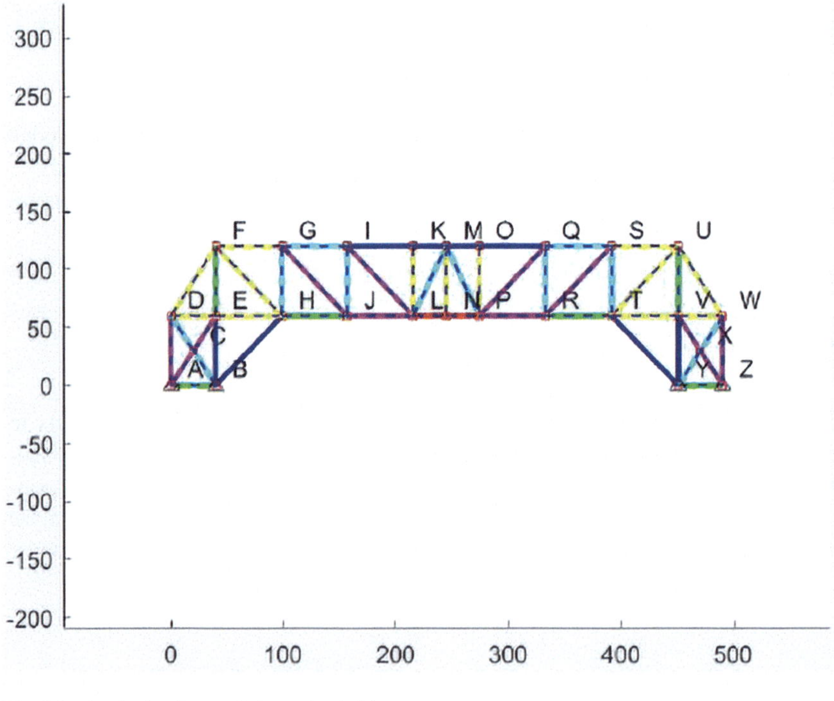

Fig. 3.3 Analysis of a model wooden bridge

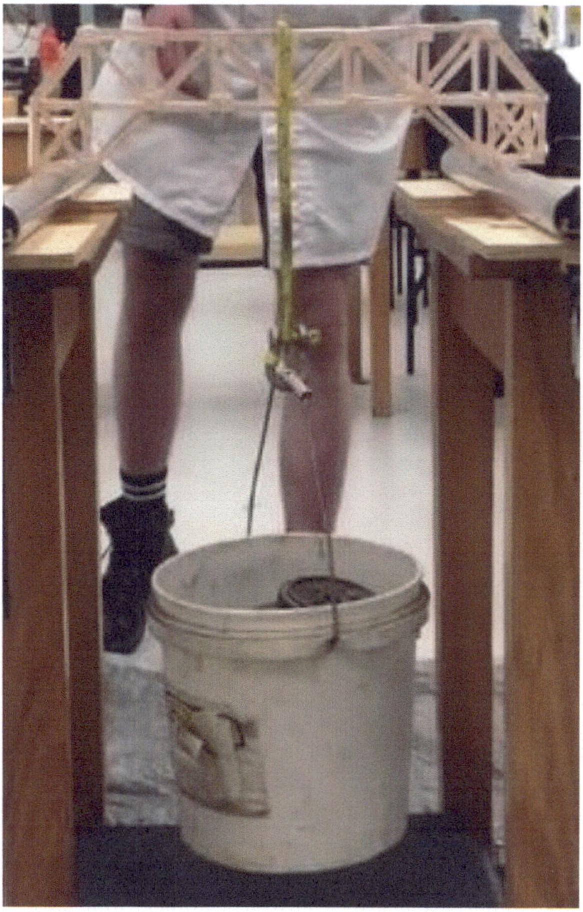

Fig. 3.4 Testing a model wooden bridge

3.9 Engineering Laboratories and Workshops

The Professor of Civil and Mechanical Engineering at the University of Tasmania, Archibald "Arch" Oliver was appointed in 1955. Professor Oliver had a vision for engineering education which was that lecture material should be built on a foundation of practical skills. As a result, the new engineering building on the Sandy Bay campus built in 1956 has generous laboratory and workshop spaces. In an unpublished manifesto entitled 'University Training of an Engineer' written soon after the completion of the Sandy Bay campus, Professor Oliver wrote about the difference between engineers and scientists, how engineers discover how to solve problems with too little or too much information, and how students can learn from laboratory experiments which don't 'work' in contrast to the way experiments in the physics department are carefully designed to prove a 'law'. The commitment to 'hands on' practical work in the curriculum at the University of Tasmania continues to be a valued component of the engineering undergraduate degree.

> Some ten years or so ago, we, at the University of Tasmania, started to face up to the problem of trying to establish what we could best do towards training engineers in the present-day environment. Technology was, and is even more so today, advancing at such a rapid rate that, by the time material was available in textbooks, it was already out of date. We were, and still are, a very small group of lecturing staff with very limited facilities so there was little chance of reading all the current technical literature and keeping up with every field of engineering development. We had to confine ourselves to a few relatively narrow lines of interest and become good at those. There were a few articles starting to appear in overseas journals expressing concern about the lack of teaching of design in university courses for engineers. We did not seem to be doing much about that either. Yet, our graduates seemed to be doing pretty much as their employers expected, and we were particularly proud that a

number of people with whom we had struggled to get up to our examination standard seemed to be doing pretty well. We wondered why.

If we were training engineers to solve engineering problems, then the problem of providing the training must also be an engineering problem and should be amenable to the same line of attack that engineers use. This was something of a circuitous argument, but it was enough to get started. When in doubt, engineers go to a code of practice and this is a summary of previous experience in similar endeavors. Although most engineers seem to jeer at the conservatism of their codes of practice, on the whole they work well, because they express the united opinion of a large number of sensible men and such opinion is usually fairly close to an optimum solution. There was no code of practice for training engineers, but we could and did look carefully at what the practice was.

Most Australian degree courses in engineering have followed the same pattern for many years. They are four-year courses, usually with two years of mathematics, one or two years of physics and one year of chemistry: the first year is nearly identical with that for the science degree and the engineering subjects are taught in the following years. Most universities had introduced early specialization particularly for electrical engineers, but the University of Tasmania still adheres to the old-fashioned idea of three years of training common to all branches of engineering with specialization in the final year only. There has been a gradual increase in the science content of all courses and a corresponding decrease in the content of purely descriptive material. All engineering courses have become more mathematical over the years, and some are starting to look very like science courses. This is particularly marked in England, but this seems to be the source of the loudest complaints about universities not teaching design. Apparently, engineers need a somewhat different training from scientists.

We then looked at what engineers did. Many young graduates were using information we had taught them, but this was mainly in design offices and most of the information they used had been taught in the early years of the course. There were many more in the construction, maintenance and production fields doing quite different things from what they had been taught to do at the university. Nearly all of the older engineers were administrators, and their work seemed to have no relationship at all to their university course work. We could find only a few who used the information taught in the final year of the course. Yet all universities had agreed that this was a proper course for training engineers and most graduates were moderately well satisfied. It was obvious that the information we handed out in lectures was not the vital thing in itself but that there was some other property of engineering courses that was providing the required training.

The next step was to identify some property of engineer's work which was common to all engineers. This is easiest to see in the design office where an engineer uses his university training almost directly; here the work often looks much like that of a scientist yet engineers state that a training in science of pure mathematics is not enough, that there is something extra in an engineer's training that makes the difference between a scientist and an engineer.

We were able to identify a number of differences between the attitudes of the average scientist and the average engineer. The scientist is more interested in finding out why, whereas an engineer has to make up his mind to get something done. Limitations of time and cost are more apparent to engineers usually - not only must he do something about his problem, but he must often do it fairly quickly and cheaply. The engineer is much more inclined to accept an approximate solution to a problem for the sake of getting on with the job whereas scientists are often severely critical of this attitude and are likely to say that the engineering attitude is rough and crude without realizing that much fine measurement is done by engineers. But there is something more basic for our needs than this: a big difference between a university engineering course and a technical college course is the emphasis on the scientific attitude at the university compared with the more direct approach of the technical college, yet university graduates are not inferior to technical college people in getting the job done.

The big difference between the work of the average scientist and that of the average engineer is that the engineer's problems are less well formulated. A scientist likes to isolate one aspect of a problem from the rest and study it separately, whereas this is usually not possible for engineers. Their problems are firmly established in the real world around them and the real people associated with them. Seldom can an engineer obtain exactly the right data about his problem. He usually has too much data regarding some aspects so that his data appears to conflict; a mathematician would say that he has more equations than he has variables in those equations, or that there is redundant data. Contrariwise the engineer has two little data about other aspects of his problem. This often applies to a very large part of his problem and he cannot afford the cost of getting more. In this sense, engineering problems are more difficult than those of scientists and an engineer has to act more like those at the frontiers of science where the larger problems are. This was the first clue, that we should be training engineers to handle large problems with redundant, conflicting and insufficient data in different problems. Apparently, the standard university course did this fairly well or it would not have become the standard, but we had still to learn how to do deliberately what we felt we were doing by accident.

We then examined in more detail just how an engineer goes about the solution of the large problems in the design office. He first collects the available data and examines it; he cross-checks his data for consistency and so obtains an estimate of its accuracy where this is possible. He then simplifies his problem. He identifies certain aspects as being most important (just as I have tried to do in this study) and sets about solving his simplified problem. For example, in the analysis of stresses in a dam the material is usually called elastic. This means that the engineer sets up a physical model of his problem or more commonly a mental picture of a physical model. We have been in the habit of calling these simplifications by the name of "assumptions" but this word implies a belief that they are they are in some way correct whereas we found that no good engineer believes that his "assumptions" are correct - they are simplifications accepted for the purpose of getting on with the job. A better description of them is given by the term "model definitions" because this is how the engineer treats them as statements used to define a model which is a simplified copy of the real problem.

The physical model is of no use in itself for design purposes until some numerical values are attached to it. This is sometimes done directly by construction of the model and testing it in the laboratories in much the same way as the scientist isolate some aspect of his problem for detailed study. More often, however, the engineer sets up a mathematical model of this physical model. Frequently this requires further "model definitions" which are necessary to make the mathematics amenable to solution so that the mathematical model differs from the physical model which one would test in the laboratories and is usually a stage further removed from the real problem.

In this context the university lecture course consists largely of setting up and solving mathematical models, but we had fallen into a number of traps which lessened the value of the training. The use of the word "assumption" has already been mentioned. Another word that had been badly used was "ideal" when we meant a mathematical model. Basically, "ideal" is not too bad a word because it contains the notion of an idea or mental image, but it is also associated with perfection, and this leads to the feeling that there is something better about the mathematical model than about the real problem. Of course, it is the other way around, the only correct model of a real problem is itself with all of its complexities. The Greeks had fallen into the same trap and made such statements as that the paths of planets must be on spheres because this was the ideal shape so that if the evidence of the senses contradicts this then the senses cannot be ideal. This trap is to be avoided at all costs. No engineer should believe that his model is anything but an imperfect copy of the real problem. Solutions for the mathematical model are just that and no more. Interpretation of these solutions in terms of the solutions for the real problem is often a difficult process. This was one feature that could be brought into our teaching to improve it.

The next step was to find out what engineering administrators do. We asked why engineering training was useful for administrators as there is no obvious connection between the work they do and what they are taught at a university. The answer given was that engineers are trained to think clearly but this was not much help to us in trying to improve their training so that they would think more clearly. We were told that engineers have logical minds and this gave us a clue because there can be no logical argument until the problem is defined whereas engineering administrators are dealing with very ill-defined problems indeed. Many of their problems deal directly with the handling of men and this is a field in which the definition is very poor.

It is apparent that engineers with their strong training in mathematics approach problems of management and administration in much the same way that they are accustomed to use in solving technical problems. They simplify the problem by ignoring those complexities which are thought to have minor effect and examine the simplified problem - a logical model replaces the mathematical model. The logical model is well defined with no redundant or conflicting information as is the mathematical model but the data consists of facts rather than numerical values.

The conclusions from the examination of the logical model are then used as a guide for making statements about the real problem. It is this same rigid logical argument that is so baffling to non-engineers such as accountants and wives. This approach would aid our teaching if we could, from time to time, point out the analogies between logical models that could be used to aid solution of these less well-defined problems and the mathematical models that are used to solve technical problems.

When we turned again to the standard Australian university engineering course, we found that it did indeed provide the sort of training required, except that there were a few drawbacks mainly arising from the use of the words "ideal" and "assumption". There was a growing tendency to think of the mathematical model as rather more real than the real problem. This is particularly apparent in the laboratory courses where it is very difficult to overcome the temptation to ensure that all laboratory experiments "work". In the terms of our models an experiment which "works" is one in which numerical test results agree with the numerical values obtained from the solution of the mathematical model, but this is bad teaching, because then the laboratory experiment is a new physical model of the mathematical model and is frequently further removed from the real problem. This is the sort of attitude that is adopted in high school science courses where the problems must be grossly simplified to be amenable to solution by the simple methods available to the high school student. Teachers usually feel more satisfaction in presenting solutions to problems which can be justified as good models by agreement with simple laboratory tests than with comparing the simple models with the real problem. The old-fashioned satisfaction of marveling at how closely a simple model can resemble a real problem is no longer popular. Seen in this light, most of the so-called "laws of nature" turn out to be either definitions, or

> axioms, which cannot be proved, such as the "laws of thermodynamics", or model definitions which define a simplified model such as "the perfect gas law". Therefore an experiment designed to "prove Hooke's law" is a badly designed experiment because the linear elastic material is only a model of a real material. Fortunately it is a good model for such materials as steel.
>
> On the other hand, we must be careful that laboratory experiments do not show up the mathematical model in such a poor light that our students are completely discouraged from using models as a guide to the solution of real problems or they would never learn to tackle a difficult problem. This is just as dangerous as presenting experiments that work so well that students obtain the idea that the mathematical model is exact and get too much respect for their models. The arguments between what are usually called theoretical and practical men arise from this source.
>
> A good engineer is one who uses models to their limit to obtain useful guides to the solution of real problems. This is the attitude we wanted to engender in our students. (Oliver, n.d.)

Despite these words being written over 60 years ago the philosophy underpinning the training of engineers at the University of Tasmania and many other Australian higher degree institutions is still relevant. With this approach to engineering education, it is not surprising the UTAS School of Engineering has produced many very able and some brilliant graduates over the years.

Engineers see the world differently from scientists and their training should reflect this difference. Whereas the scientist seeks to understand the reason for what is observed in nature, the engineer is more concerned with providing an outcome in the face of nature's challenges.

3.10 Assessment Tasks

Assessment tasks are fundamental to assuring graduates' competencies. In earlier times competence was assessed directly; for example, an apprentice blacksmith would demonstrate the ability to shoe a horse, before acceptance as a journeyman in the trade. Technical and Further Education (TAFE) and Vocational Engineering and Training (VET) colleges have adopted the same competency-based assessment model. A student is assessed to be either competent or not competent. Universities have traditionally relied on formal, invigilated, written examinations to assess the level of a student's knowledge, skills and understanding. The assessment may be graded (according to mark) as fail, pass, credit, distinction, or high distinction.

With the adoption of Constructive Alignment (Biggs et al., 2022) assessment tasks are aligned both to the Intended Learning Outcome (ILO) of a subject as well as the content delivered in the subject. To give an example from a first-year engineering subject—the ILO is 'to understand the design of simple truss bridges'; the content delivered comprises lectures/tutorials on calculating forces in the members of a truss bridge for a particular applied load followed by 6 h in a laboratory/workshop fabricating the model bridge from balsawood and glue. The assessment task is to write a report on the design, fabrication and testing the model bridge. (Fig. 3.2). This type of assessment is described as a 'summative'. The mark given to the report is part of the overall assessment in this unit.

There may also be 'formative' assessment tasks which do not contribute to the overall grade—for example submitting a preliminary design of the model bridge to the lecturer/tutor for feedback. As well, some units may require a 'hurdle task', usually scheduled in the first week or two, to assure a student's enrollment, which is assessed as pass or fail but does not contribute to the overall grade.

Depending on the nature of the subject there is a range of assessment methods. These may include assessment of a student's knowledge through examination which may, or may not be, of fixed duration (e.g. 3 h or 24 h); open book (textbooks and/or lecture notes permitted); written, computer-based or oral. Written reports (either individual or group) are the usual form of assessing laboratory or field work. Students may be required to give an oral presentation of their work to an audience of fellow students and lecturers.

3.11 Marking Rubrics

Each assessment task is required have a marking rubric which is made available to students.

Table 3.2 is a marking rubric for a second-year unit at UTAS 'Engineering Design and Build' in which students were challenged to build a robot to navigate a pathway and deliver tennis balls at different locations. This rubric has components specifically related to mechanical and electrical elements of the design as well as the management of the project.

3.12 Academic Integrity

Cheating in examinations is an issue for meaningful assessment. Cheating can take many forms from simply copying another student's work to contracting another person to write a report or provide answers to examination questions in real time. Students have been caught passing notes in invigilated examinations, even signing answers to others with hand-signals.

In one case a student was found to have completed all questions 15 min after the start of a 3-h examination. When challenged, the student asserted he was a fast writer. However, on a close scrutiny the pre-printed examination script was subtly different from others handed out; the staple was in a different position and the grey areas were slightly darker. The student clearly had obtained a copy, completed the script and had smuggled it into the examination room. This situation posed a problem for the assessor—the obvious remedy was to fail the student on the basis of the

Table 3.2 Marking rubric for second-year unit

ENG201 rubric Student Name and Number

Submission should be as normal assignment (PDF) on MyLO, but may be supplemented by videos (submitted with the report on MyLO or links to online copies eg YouTube, Vimeo, etc)

Length is as per the unit outline - 10 pages for the main report plus an appendix (if desired).

	Component (these should be the major headings in your report)	Weight	Mark Awarded	HD	DN	CR	PP	NN (40-49)	NN (0-39)
Your INDIVIDUAL contributions	**1. Contributions** **Individual Design Implementation Contributions:** Show/describe your individual contributions to the implementation and testing of the design. For example: - mechanical manufacture, fabrication, testing and refinement - algorithm development, microcontroller coding, electrical system testing and "case hardening". **Individual Project Management Contributions:** Show/describe your individual contributions and/or participation in the management of the project and the team. **2. Associated Enhancement of your EA Stage 1 Competencies** Reflect on how making the above contributions and learning experiences have enhanced your EA Stage 1 Competencies, providing specific examples and references to stage 1 competencies.	50.0%		Individual contributions demonstrate an exceptional level of engagement and contribution that have made outstanding contributions to the development, functionality and reliability of the design. Comprehensive evidence of the contributions has been provided by way of description, diagrams, photos, etc. A thorough and elegant description provided of individual contributions/participation in the management of the project and the team demonstrating critically valuable input that helped achieve project goals. Very convincing evidence of comprehensive and extended enhancement of EA Stage 1 Competencies referencing specific stage 1 competencies in each of the categories of "Knowledge and Skills Base", "Engineering Application Ability", and "Professional and Personal Attributes".	Individual contributions demonstrate an high level of engagement and contribution that have made valuable contributions to the development, functionality and reliability of the design. Good evidence of the contributions has been provided by way of description, diagrams, photos, etc. A good description provided of individual contributions/participation in the management of the project and the team demonstrating valuable input that helped achieve project goals. Convincing evidence of good enhancement of EA Stage 1 Competencies referencing specific stage 1 competencies in each of the categories of "Knowledge and Skills Base", "Engineering Application Ability", and "Professional and Personal Attributes".	Individual contributions demonstrate engagement and contribution that have made contributions to the development, functionality and reliability of the design. Evidence of the contributions has been provided by way of description, diagrams, photos, etc. A description provided of individual contributions/participation in the management of the project and the team demonstrating input that helped achieve project goals. Evidence of enhancement of EA Stage 1 Competencies referencing specific stage 1 competencies in each of the categories of "Knowledge and Skills Base", "Engineering Application Ability", and "Professional and Personal Attributes".	Individual contributions made towards the development, functionality and reliability of the design. Evidence of the contributions has been provided. A description provided of individual contributions/participation in the management of the project and the team demonstrating some useful input. Evidence of enhancement of specific EA Stage 1 Competencies.	Few individual contributions made towards the development, functionality and reliability of the design. Little evidence of the contributions has been provided. Few individual contributions/participation in the management of the project and the team demonstrating useful input. Little evidence of enhancement of specific EA Stage 1 Competencies.	Individual contributionstowards the development, functionality and reliability of the design have not been made or are substantially lacking. Evidence of the contributions has not been provided or is substantially lacking. Individual contributions/participation in the management of the project and the team have not been made or are substantially lacking. Evidence of enhancement of specific EA Stage 1 Competencies has not beengiven or is substantially lacking.
Your review of the TEAM project	**3. Review of your team's design and its performance** Review your team's design outcome. Critically analyse and reflect on the design and its performance. Describe/show/explain how your design performed in testing and in competition and in comparison to what was expected. Discuss what aspects of the design functioned well and what/how it could be improved. Include all relevant mechanical and electrical aspects. Illustration with video is encouraged (submit the videos on MyLO or provide links to online videos).	25.0%		Demonstrates excellence in design and implementation, evidenced by a comprehensive review and critical analysis of the design and its performance in relation to each aspect of the desired functionality and competition. Key features are highlighted showing the design has been developed to a very high standard. Excellent performance of the design is shown, evidenced by performance in testing and competition. Potential improvements that will optimise the design and performance are identified.	Demonstrates a high standard of design and implementation, evidenced by a good review and critical analysis of the design and its performance in relation to each aspect of the desired functionality and competition. Key features are highlighted showing the design has been developed to a high standard. Very good performance of the design is shown, evidenced by performance in testing and competition. Potential improvements to the design are identified.	Demonstrates good design and implementation, evidenced by a review and analysis of the design and its performance in relation to aspects of the desired functionality and competition. Key features are highlighted showing the design has been developed to a good standard. Good performance of the design is shown, evidenced by performance in testing and competition.	Demonstrates a satisfactory level design and implementation, evidenced by a review of the design and its performance. Key features are highlighted showing the design has been developed to a satisfactory standard. Some aspects of performance of the design are shown.	Does not demonstrate satisfactory design and implementation from the evidence provided of the design and its performance. Few key features have been developed to a satisfactory standard. The performance of the design is poor.	Design and implementation has progressed little from the concept design stage, based on the evidence provided of the design and its performance.
	4. Project and Team Management Reflect on the design outcome in relation to the management of your project. Include aspects such as how your team functioned, team contributions in terms of "Belbin Team Roles", things that went well and those that didn't, and also how the project and team management could have been improved.	25.0%		Demonstrates excellence in critical analysis and reflection on the design outcome in relation to project and team management. Insightful reflection made on management of the project and how the team functioned. Specific team contributions appropriately critically discussed in terms of "Belbin Team Roles" and comprehensively illustrated with relevant examples. All aspects of project and team management reviewed and appropriate suggestions for improvement made.	Demonstrates a high standard of critical analysis and reflection on the design outcome in relation to project and team management. Reflection made on management of the project and how the team functioned. Specific team contributions critically discussed in terms of "Belbin Team Roles" and illustrated with examples. Most aspects of project and team management reviewed and appropriate suggestions for improvement made.	Demonstrates a good standard of critical analysis and reflection on the design outcome in relation to project and team management. Reflection made on management of the project and how the team functioned. Specific team contributions discussed in terms of "Belbin Team Roles". Some aspects of project and team management reviewed and appropriate suggestions for improvement made.	Demonstrates some critical analysis and reflection on the design outcome in relation to project and team management. Reflection made on management of the project and how the team functioned.	Demonstrates little critical analysis and reflection on the design outcome in relation to project and team management or on how the team functioned.	Critical analysis and reflection on the design outcome in relationto project and team management or on how the team functioned has not been made or is substantially lacking.

Notes:

Comprehensive evidence means multiple examples and a wide range of good quality evidence. Note that, in general, it usually takes three pieces of evidence to convince someone that what you are telling them is true! You should therefore make use of descriptions illustrated by diagrams, photos, videos, etc., wherever possible. It would depend on the purpose of the evidence and the nature of the photos and videos whether they would be enough – they sometimes are, but not always.

For this report the information sheet and rubric set out what you need to achieve, and include some guidance. However, it is part of the assessment task for you to decide how best to do this and what to include. For example, in reflecting on the management of the project, if you conclude that you did not allow sufficient time for various aspects of the project, then you may want to show what should have been done, for example, by presenting a revised Gantt chart showing suggested improvements over an original chart. Another example is if you didn't prepare individual weekly work schedules for each team member and now think that would have been a good idea, then you could show suggested schedules. It will help to reflect on aspects of project management that were covered in the lectures and experienced in the project complemented by specific supporting material whenever possible.

evidence of cheating despite protestations of innocence. The solution was to invite the student to re-write the examination under supervision.

With the advent of on-line examinations cheating is difficult to detect. Students are warned that cheating and plagiarism will not be tolerated; they are instructed on the accepted methods of attributing the work of others; written reports are routinely scanned by plagiarism-detecting software such as 'Turnitin'. Penalties for cheating or plagiarism can be as severe as exclusion from a course.

Late in 2022 generative artificial intelligence (GenAI) became readily accessible. Students were able to generate technical reports of a very high standard, although not for reports of practical work in which the results, particularly for engineering experiments, are unpredictable. The following is advice to students at the University of Tasmania on using GenAI (Davidson & Boyes, 2023):

Using artificial intelligence AI for study.

- You can use generative artificial intelligence AI to learn just like you would study with a classmate or ask a friend for advice.
- You are not permitted to present the output of generative AI as your own work for your assignments or other assessment tasks this constitutes an academic integrity breach.
- In some units a unit coordinator may explicitly allow or require the use of AI in your assessment task.

3.13 Concerns

As a result of the COVID-19 pandemic there was a rapid uptake of on-line delivery with lecturing staff and ICT technicians scrambling to facilitate lectures on Zoom. The shutdowns forced universities to deliver all content on-line. When classes recommenced in February 2020 most students were able to access the on-line content although for students unable to return to Australia from overseas their internet bandwidth limitation initially precluded visual contact.

In the later months of the pandemic students were reluctant to return to face-to-face delivery. This was partly from students becoming comfortable with off-campus study and the opportunities afforded for part-time work, but also because university administrators discovered efficiencies in space utilization (not needing large lecture theatres) and staff productivity (recycling recorded lectures).

3.14 Blended-Learning

The outcome, post COVID, has been mandating blended-learning with content material posted on-line. Face-to-face is restricted to tutorials, workshops and laboratory sessions. The pre-COVID style, 50-min presentations, to a lecture theatre full of students have disappeared.

The production of high-quality on-line learning materials is time consuming and requires special skills. Modern students can view material prepared by professional educational designers. In a small regional university like the University of Tasmania the responsibility for the production of online materials material falls on the lecturer. There is limited access to help from educational designers.

In the past, a lecturer would deliver content material face-to-face to a lecture theatre full of students. Thelecturer would write the material on the blackboard, draw diagrams freehand and write the mathematics while explaining and engaging with students on the topic. At the end of 50 minutes each student would have a handwritten record of the lecture, they would have heard the lecturer's explanation and would have had the opportunity to query any of the content and receive clarification. Preparing the same material with print-standard diagrams and mathematical symbology might take a lecturer perhaps 5 or more hours and would not look as good as the professionally prepared material from the larger, better resourced, universities. Besides, students accessing these materials on-line do not have the opportunity to engage with the lecturer.

With blended-learning, students are expected to study the online materials before attending face-to-face tutorials or workshops. The success of this method relies on the students having the time and motivation to access the online material beforehand. In practice, time is wasted at the face-to-face sessions revising the on-line material for students who have not viewed it. This limits the time for engaging with students and concentrating on those threshold concepts which underpin the subject matter. There is, however, the advantage for some students in being able to speed up or slow down an on-line presentation.

3.15 Teaching Versus Research

The modern academic is contacted to spend 40% of their time teaching, 40% on research and 20% on administration and community activities. The preparation of online materials, as explained above, is at the expense of their other duties. Their research output suffers and consequently their opportunity for promotion. Universities have the option of engaging 'teaching focussed' staff who are required to devote 80% of their time and 20% to administration and community activities. These staff have limited opportunities for promotion within the university system which is essentially research driven.

There is a danger, from the point of view training engineers, in lecturers preferencing their research at the expense of quality teaching and learning. Ideally, engineering academics should have the time and opportunity to engage with industry to maintain their currency. Ideally, also, a lecturer's research and industrial currency should inform their teaching. Much of the undergraduate curriculum is far removed from the cutting edge of research. Students in a small university are fortunate indeed to learn from an academic with the opportunity to weave their research into their teaching.

References

Engineers Australia. (n.d.). *Stage 1 competency standard for professional engineer.* Retrieved 24.10.23 from https://www.engineersaustralia.org.au/sites/default/files/2022-07/stage-1-competency-standard-professional-engineer.pdf

Biggs, J. (2014). Constructive alignment in university teaching. *HERDSA Review of Higher Education, 1*, 5–22.

Biggs, J., & Tang, C. (2011). *Teaching for quality learning at university*. McGraw-Hill Education.

Biggs, J., Tang, C., & Kennedy, G. (2022). *Ebook: Teaching for quality learning at university 5e*. McGraw-Hill Education.

Carew, A.L, Doe, P., Hadgraft, R., Symes, M. and Henderson, A. (2013) Engineering curriculum structure and mapping: accreditation and beyond. In *Proceedings 2013 AAEE Conference. Gold Coast, Australia.*

Davidson, S., & Boyes, A. (2023). *Artificial intelligence and academic integrity*. Retrieved from University of Tasmania, Academic Division. https://universitytasmania.sharepoint.com/:p:/r/sites/academic-division/_layouts/15/Doc.aspx?sourcedoc=%7B599FA5AE-0DA8-4EF6-A5D4-5781398BE551%7D&file=Presentation%20-%20Artificial%20Intelligence%20and%20Academic%20Integrity%20September%202023.pptx&action=edit&mobileredirect=true&DefaultItemOpen=1

Doe, P. E. (1998). *Valedictory speech*. Institution of Engineers Australia. (unpublished).

Holmes, D., Sheehan, M., Birks, M., & Smithson, J. (2018). Development of a competency mapping tool for undergraduate professional engineering programs, using mechanical engineering as a case study. *European Journal of Engineering Education, 43*(1), 126–143. Retrieved from : https://eprints.qut.edu.au/223075/ 05.11.23.

Oliver, A. R. (n.d.). *The university training of an engineer*. University of Tasmania unpublished.

SoE. (2013). *Engineers Australia accreditation*. University of Tasmania School of Engineering (unpublished).

Chapter 4
UTAS Faculty of Engineering, 1894–2023

Abstract The fourth chapter chronicles the history of the Faculty of Engineering at the University of Tasmania. It recounts some of the twists and turns and the role of personalities and ambitions in evolving from a "Faculty of Engineering" to a combined "Department of Engineering and ICT" and the "divorce" when ICT (information and communications technology) emerged as a separate department. It is a personal account from the author of this book who was the head of the Department of Engineering in Tasmania's second city, Launceston, between 1996 and 2000. The "1980 rule" saw a shift from 2- to 3-year engineering certificate and diploma in the TAFE sector to Bachelor of Engineering degrees awarded by the newly formed Tasmania College of Advanced Education.

4.1 Introduction

The rationale of this chapter is both to chronicle the establishment and development of one of the first Australian schools of professional engineering and also to provide an insight into the University of Tasmania's responses to directives from the Australian Government, the prime funding body for Australian universities, and the Institution of Engineers Australia (now renamed as Engineers Australia) which is the national body accrediting engineering degree courses. The author of this book was for a time a head of department within the UTAS School of Engineering and Architecture and relates from his experience how decision-making within the faculty was influenced by competition between departmental heads for funding and the perceived importance of their disciplines.

The University of Tasmania's School of Engineering was founded in 1921. Until the establishment of the Australian Maritime College in 1990, the University of Tasmania was the only provider of professional engineering education in the state. At present, the School of Engineering provides courses leading to bachelor's degrees in civil engineering, mechanical engineering, electrical power engineering, electronics and communications engineering, and computer systems engineering. A

P. E. Doe, *Prospects for Professional Engineering Education in the Asia-Pacific Region*, Education in the Asia-Pacific Region: Issues, Concerns and Prospects 74,
https://doi.org/10.1007/978-981-96-3197-1_4

new specialization, mechatronics engineering, was introduced this year. The Australian Maritime College, which is affiliated with UTAS, offers courses leading to bachelor's degrees in naval architecture, ocean engineering, and marine and offshore systems.

Prior to the formation of the UTAS Faculty of Engineering, there had been a School of Mines and Metallurgy in Zeehan on the West Coast of Tasmania since 1892 (Fig. 4.1). The West Coast of Tasmania at that time was experiencing a minerals boom with the copper mine and smelter at nearby Queenstown and a gold, silver, lead, and zinc mine in Rosebery.

The course at the Zeehan School of Mines could be considered to be delivering a professional engineering qualification in that it included mathematics, chemistry, mechanics, mechanical drawing, and surveying.

Minutes from a meeting of the board of the School of Mines and Metallurgy dated May 14, 1896, reported the following item from staff members:

> As lecturers to the Zeehan School of Mines we beg to submit the following report for the term now concluding.
>
> - During the term classes have been held regularly morning and evening in the following subjects: Mineralogy and blowpiping, chemistry, assaying, mathematics, mechanics, and mechanical drawing.
> - The total number of students attending the school has been 24.

Fig. 4.1 The Zeehan School of Mines (Chalmers & Meikle, 2017a)

- The number in each subject has been as follows:- Mineralogy 10, chemistry 7, assaying 5, mathematics 12, mechanics 10, mechanical drawing 15.
- The attendance has been on the whole very good. The following figures give the average attendance in each class - Mineralogy 7.4, chemistry 5.6, assaying 4.1, mathematics 8, mechanics 8.8, mechanical drawing 9.11. (Zeehan School of Mines, 1896)

The School of Mines trained engineers and managers for the local and interstate mining industry until the 1920s. Its history is recorded in the book compiled to celebrate the 125th anniversary of the University of Tasmania:

> The first 10 years in the University of Tasmania's history were turbulent to say the least. Tasmania was in a long period of recession, and many Tasmanians saw the University as an extravagance that the state could not afford. Many members of Parliament went so far as to say the University should be closed, and the funds diverted to the state's floundering primary education system. This level of hostility was particularly strong in the north of the state, with local MPs portraying the University as a waste of money that didn't benefit their electorates, despite the University's best efforts to work with them.
>
> Parliament, press and people alike were all determined to write-off the University as unnecessary. This led the University board to propose a new School of Mining Engineering. By opening a school that so obviously benefited the people of Tasmania, they hoped it would put to rest any doubts about the University's value to the state. Vice-Chancellor Walker envisioned scientific training at the University, followed by practical training at the existing Zeehan School of Mining on the West Coast, with the University maintaining international academic standards.
>
> Unfortunately, opening a new school meant more government funding, and during the recession the state government had been less than generous with its tax-dollars. This case was no different, as the initial proposal for £600 (equivalent to $79,000 in 2013) was rejected. Eventually, however, the University was able to secure a grant of £300.
>
> The public was also less than supportive of the idea. The Hobart-based right-wing newspaper *The Clipper* was extremely critical of the proposal, saying that the government was wasting money on the Zeehan School of Mines so that "a few fat men's sons may have their degrees confirmed by a professor of dead languages."
>
> Despite the lack of support, the University pushed ahead with its plan for a new mining school. However, they soon ran into difficulties in 1901, when not a single student signed up for the new course. The newly appointed Mining Engineering Lecturer Robert Smith lamented that "The public did not as yet appear to be alive to the advantages provided by the course." Later in the year, Professor Smith was able to obtain five students to instruct, a reasonable number for a new course at the time.
>
> At the turn of the century, the University had enjoyed some support from the Lewis Administration. Sir Neil Elliott Lewis was previously a member of the University Council before he became Premier, and although his administration was often constrained in the face of public opinion and hostility from other MPs, they did what they could to assist the University where possible. This security abruptly vanished in the 1903 elections, when the Lewis government was swept from office by a much more radical Labor administration led by William Propsting.

With this shift in power, it wasn't long before the University was once again in the firing line. In the north of the state, denouncing the "southern university" had almost become an annual tradition for local MPs. This time, however, the focus was on the new School of Mining. Devonport member John McCall lamented the further waste of taxpayers' dollars, while Woollnough wished to eliminate the University's mining course, which he considered less practical than the courses offered at the Zeehan School.

Around the same time, Vice-Chancellor James Walker suddenly passed away from pneumonia. Walker had long been a strong advocate for the continuing survival of the University, and his death came as a devastating blow for the University staff. When Clarke forlornly made the announcement of Walkers death to the council, he noted that "their distress went very close to dismay." This was only made worse when the new Vice-Chancellor, Thomas Stephens, resigned after less than a year. However, his surprising successor was none other than the recently defeated Premier, Sir Neil Elliott Lewis, who quickly put his six years in the political wilderness to work defending the University from threats of closure.

This new leadership couldn't have come at a better time, when in October 1903, the state parliament once again moved to vote on the closure of the University of Tasmania. The Minister for Education, Herbert Nicholls, made a powerful speech lambasting what he referred to as the annual attack on the University. He pointed out that a third of degree candidates from the previous year were from the north of the state, that the University gave all Tasmanians access to tertiary education – rather than just those wealthy enough to travel outside the state – and that it prepared Tasmanians for jobs that would otherwise go to more educated graduates from the mainland or overseas.

These arguments were all but ignored as the states northern MPs repeated the same tired arguments they had relied on for over a decade: that the University was a waste of taxpayers' dollars otherwise needed for primary education, and served no purpose other than that of the southern socio-political elite. C.J. Mackenzie stated that abolition of the University as the best means of ending uncertainty about its future. While John Hope stated that the demise of the University was the only pledge he had given to his electorate.

Henry Dumbleton of Devonport was more original in his statement that it was more beneficial to send student's interstate, away from family ties that would otherwise be a distraction from their studies.

Despite their protests, however, the House of Assembly eventually voted against abolition of the University of Tasmania, in a vote of fifteen to thirteen. Surprisingly, the leaders of the new Labor Party generally acted as arbitrator in the debate. Everyone involved expected the Labor Party to take the same anti-university position as radical right-wing publications like *The Clipper*. If that had been the case, the University would have almost certainly faced closure. Instead, the first parliamentary representatives of the Tasmanian working-class confronted the issue in a principled and responsible manner.

But the leaders of the Labor Party did not intend to remain completely neutral, and their support of the University came at a cost. They gave notice of a tough motion to end the mining school "farce" by transferring £500 (equivalent to $67,000 in 2013) from the University to the Zeehan School of Mines. This served as the final nail in the University Mining School's coffin in the minds of both the public and University staff.

In 1904, the Bachelor of Mining Engineering was officially terminated and the funding redistributed into the Faculty of Science, allowing the commencement of new courses like mechanical drawing, surveying and applied mechanics. Professor Neil Smith anticipated the abolition of his course and had already resigned. His junior lecturer, Mr John Hilton Mackay, was much luckier. Redeployed in the revised Science Department, amongst other courses, he was able to retain his engineering title and went on to become the University's first Professor of Engineering.

Ultimately, only two students were awarded the Bachelor of Mining Engineering, one of whom acquired it as a double degree. When the University received its royal charter in 1914, the Bachelor of Mining Engineering was excluded from the degrees securing Commonwealth recognition.

4.2 The Mining School's Legacy

Some good did come out of the Mining School incident of 1902-1903. Using the grant money from the School of Mining Engineering, the University was able to make improvements in the Faculty of Science, and it also led to a new emphasis on lecturers working full-time at the University, rather than having to split their time between multiple institutions.

Student numbers also increased dramatically after the mining school incident, going from 57 in 1904 to just over 100 in 1906, and then 147 in 1910. The appointment of a former premiere as the Vice-Chancellor surely helped to give the fledging University more legitimacy in the eyes of the public, and when Lewis was re-elected as premiere in 1909 it seemed clear that the University's future in the state was secure.

It also served as a lesson for the state government. Soon after the closure of the mining school, the newly elected Labor Party set about overhauling the state's primary education system. They soon discovered that many of the state's primary school teachers had little or no formal training; which led to them being supplanted by more qualified teachers from South Australia.

This resulted in a public backlash against the government, which was so severe that it led to the resignation of the Director of Education and served to solidify in their minds the need for a university in Tasmania in order to stop specialist jobs being lost to university graduates from the mainland.

Shortly after the closure of the Mining School, the state government established the new Teachers Training College in Hobart, which worked closely with the University of Tasmania to give comprehensive training and education to aspiring Tasmanian teachers. This ushered in a new era of cooperation between the Government and the University that continues to this day. (Chalmers & Meikle, 2017a)

4.3 North-South Rivalry

The rivalry between North and South of Tasmania would color the development of the university for many years to come. It dates from the early days of the colony when Hobart was settled from Sydney in 1803 and Launceston was settled from

Melbourne a year later. Until 1810 there were two lieutenant governors, one in Hobart and one in Launceston: towns only 200 km apart. This rivalry came to a head a century and a half later with the establishment of the Tasmanian State Institute of Technology (TSIT) with headquarters in Launceston.

The vice-chancellor of the university, Professor Alec Lazenby, writes of this time:

> On hearing that I was going to the University of Tasmania a friend of mine said to me, 'the state is a bit like ancient Gaul divided into three parts but there is a difference- the divisions in Tasmania are much deeper'. Whatever the truth of this assertion, one of the more obvious challenges facing the University of Tasmania in the early 1980s, namely, to do something about higher education, had relevance to all parts of the state. Government decisions made earlier had resulted in a system that was neither rational nor efficient. Some subjects, such as pharmacy and surveying, normally provided by Colleges of Advanced Education, were part of the University in Hobart, whilst others e.g., education and some arts and science courses, were available at both the University and the Tasmanian State Institute of Technology [TSIT] in Launceston. Thus, instead of complementing each other, the two institutions were competing aggressively for a limited number of students, adding to the North/South divide in the state. Yet, Tasmania simply didn't have the population to support more than one higher education institution that was properly resourced both to provide the personnel required to meet the state's professional needs and to remain competitive with other universities.
>
> This conclusion wasn't new. David Caro, my predecessor, had tried, unsuccessfully, to effect an amalgamation of the two institutions, an objective that seemed to me to be unattainable unless there was some intervention from the Federal Government. Fortunately, the paper proposing a Unified Higher Education system, including the merging of Australian Universities and Colleges of Advanced Education, released in the mid-1980s by John Dawkins, the Federal Minister for Education, provided the opportunity for renewed discussions on amalgamation. Before this could happen, both the UTAS Professorial Board (PB) and the Council had to be convinced that merging with TSIT was in the University's interest. This requirement was only met after many lengthy discussions and much lobbying by a number of us favouring amalgamation, with two late university colleagues, Peter Byers and Phillip Hughes, playing particularly important roles in this campaign.
>
> Three members of UTAS (the Chancellor, Sir Guy Green, the PB Chairman, Phillip Hughes, and I) were then chosen to represent the University in the merger discussions. These negotiations, which began in 1987, were both difficult and protracted. Initially, the Chairman of the TSIT Council, the late Edmund Rouse, and the Principal, Coleman O'Flaherty, were unenthusiastic about giving any serious consideration to merging with the University of Tasmania. When pressure from the Federal Government made such an amalgamation almost inevitable, it is understandable that the TSIT negotiators wanted any agreement to embody the best possible outcome for their institution. They were thus strongly in favour of a university with a federal, rather than a unified, structure; felt that the newly merged institution should not retain the name 'University of Tasmania'; and argued that the office of the Vice-Chancellor should not be on, or adjacent to, the campus of either participating institution.
>
> These proposals ran completely counter to my ideas of the newly merged institution. I had envisioned that the amalgamation of UTAS and TSIT would provide the opportunity to establish an institution widely accepted as the university serving the whole of Tasmania; this institution could also make a significant contribution to greater unity within the state. The only logical decision would therefore be to retain the name 'The University of Tasmania'. The necessary visible presence in all three regions seemed achievable without too much difficulty. Established campuses already existed in Hobart and Launceston whilst

the University had recently set-up a presence in Burnie. This centre, established to provide some first-year university courses to help increase the participation rate in the Northwest, could provide a possible site for developing a campus in the region.

The realisation of these objectives required a number of things to emerge from the negotiations. First, any amalgamation should be a genuine merger of the expertise and other resources of UTAS and TSIT, rather than being perceived as a "takeover" of the one institution by the other. Secondly, in order to effect the proper and efficient use of these resources, a structurally unified university was essential. In Tasmania especially, with its history of division and conflict, particularly between the North and the South, a federal university would simply provide an opportunity for the financially autonomous components to continue to make decisions based in their own interests rather than the needs of the state as a whole. Further, I could not imagine any circumstances where the Vice-Chancellor's office should be completely separated from any of the university's teaching and research activities.

Fortunately for the negotiators, external factors played a big part in determining both the structure and the name of the new university. John Dawkins attended one of our merger meetings and made clear that his idea of an amalgamation was the creation of a unified institution. In addition, the naming of the new University was taken out of the hands of the negotiating body when a state Minister, Peter Patmore, proposed that it should be called 'The University of Tasmania', a proposition accepted by the Tasmanian parliament.

The negotiating committee's recommendation for the merger of UTAS and TSIT was subject to agreement of the draft Parliamentary Act. Predictably, there were many lengthy discussions on the wording of this document, especially on some contentious issues. In spite of matters being determined by a clear majority only after in-depth deliberations, some decisions were not accepted by everyone. I particularly remember one critical Council meeting when the Chancellor ruled that two of these issues – the structure of the new university and the composition of its Council – had been traversed sufficiently and disallowed further discussions on them. The act was then debated and ultimately passed by Parliament. The University Council received the first notification of this decision in the form of a fax sent to me by Ross Ginn [the Legislative Council's representative and a strong supporter of the amalgamation] during my last Council meeting. The fax read, 'Baby safely delivered at 3.15 this afternoon'.

I had just one regret about the merger discussions, namely that we failed to persuade the Australian Maritime College [AMC] to become part of the new university. The College argued the need for financial autonomy, a proposition that had some legitimacy because of its specific and discrete program. However, the merger discussions with the AMC occurred at the same time as those between UTAS and TSIT. This was unfortunate because accepting the AMC's demand for financial autonomy then would almost certainly have been taken as a precedent that would have resulted in disastrous implications for the structure of the new university.

I had decided to retire when the newly merged university, with its unified structure, had been accepted by government. I felt strongly that its first VC should come from outside the merged institutions, thus removing any perception of any bias in the implementation of the Act that might arise from the appointment of an internal candidate. The initial responsibility for this task, just as important as the wording of the Act in determining just how successful the new institution would become, fell to my successor, the late Alan Gilbert. No one could have been more effective than he was in ensuring that the proper foundations of the new University of Tasmania were so firmly laid.

About the author: Professor Alec Lazenby AO, LLD (Honoris causa) (1991), was formerly Professor of Agronomy and Vice-Chancellor, University of New England and

> Director of the Grassland Research Institute (UK). He was Vice Chancellor of the University of Tasmania during negotiations between the University and the Tasmanian State Institute of Technology (TSIT) which led to the merger of the two institutions, promulgated on 1 January 1991 as the present University of Tasmania. (Chalmers & Meikle, 2017b)

The amendment to the act of the Parliament of Tasmania, in 1972, deleting a clause restricting the university's Sandy Bay campus to be used for "educational purposes," was to play a significant role 40 years later when a highly contentious decision was made by the UTAS Council to abandon the Sandy Bay campus and move into the Hobart CBD.

4.4 Rationale for a Professional Engineering School in Tasmania

It has been argued in the past that engineering education in Australia-wide could be rationalized by limiting the number of universities offering Bachelor of Engineering courses as is the case for medicine and dentistry and engineering in New Zealand. The case for Tasmania having its own engineering school was well put at a function to mark the 40th anniversary of the Tasmanian consulting engineering firm Johnson McGee and Gandy (JMC). The Premier of Tasmania Mr. Bacon was reported as saying:

> "Importantly JMC has provided Tasmanian engineering graduates with work. All four directors of the company are graduates from the University of Tasmania while fourteen of the company's 15 qualified engineers are also local graduates." (The Mercury newspaper Hobart 2nd of March 2001)

4.5 The University of Tasmania's Faculty of Engineering

Engineering education at the University of Tasmania can date its genesis to the earliest days of the university with the appointment of John Hilton Mackay as lecturer in engineering within the Faculty of Science in 1903 and subsequently professor of engineering in 1920. Alan Burn was appointed as acting professor of engineering in 1919 and was professor of engineering until his retirement in 1955.

The UTAS Faculty of Engineering was established with the retirement of Professor Mackey in 1922 and the appointment of Alan Burn as professor of engineering. At that time, the physical and personnel resources of the university were located on the Queens Domain within 500 m of the Hobart GPO.

Engineering education in Tasmania was administered by an "Engineering Board of Management" which had responsibility for coordinating the university degree courses and the more practically oriented Technical College Diploma of Engineering. Both the university graduates and the technical college diplomates qualified (with appropriate experience) for admission to full corporate membership of the Institution of Engineers, Australia, at that time.

4.6 The Sandy Bay Campus

The university was to outgrow its Domain campus. In 1956, when the author of this book enrolled in the first year of a joint Bachelor of Science and Bachelor of Engineering degree, classes were held both in the Domain campus and in the drawing office, workshop, and laboratories of the Hobart Technical College across the road which was later to become the Brooker Highway.

At a meeting of the UTAS Council on February 22, 1955, the Vice-Chancellor, Professor Torleiv Hytten, reported:

> At its November meeting Council asked me to prepare a report on the probable cost of the new University Buildings at Sandy Bay. And to formulate a plan for the removal of the rest of the Faculties to that site, 'in temporary buildings if necessary'. I have ruled out temporary buildings, as it does not appear likely that the Government will consent to spending money on more temporary buildings. (UTAS, 1955)

Classes in my second year in 1957 were split between the two campuses with lectures in physics and mathematics delivered in temporary buildings on the Sandy Bay campus and "Mechanical Engineering 1" lectures in the newly erected Engineering Laboratory Building (Figs. 4.2 and 4.3). A 20-min midmorning break between lectures was time enough for students to travel by bus from the Domain campus to the new campus at Sandy Bay. Life as an engineering student at that time

Fig. 4.2 Architect's model of the engineering building on the Sandy Bay campus

Fig. 4.3 Engineering building on Sandy Bay campus c. 1965

is described in detail by Mr. Peter Watt who was later to return to his *alma mater* as a lecturer and later senior lecturer in electrical engineering.

4.7 Life As an Engineering Student in the 1950s

Peter Watt graduated from the School of Engineering at the University of Tasmania with a Bachelor of Science in 1956 and an Engineering degree with First Class

Honors in 1958. After a period as a research fellow at the National Research Council in Ottawa, Canada, working on the measurement of gravity at sea, he returned to the University of Tasmania, as a senior research fellow, and was appointed to a lectureship in 1964. Peter was promoted to senior lecturer in 1968, a position he held until his retirement in 2003. However, he continued working as a casual lecturer with the title honorary research associate in the School of Engineering until just days before his unexpected death.

Peter was highly respected for his expertise as an electronics engineer in a career that began before computers when electronics was based on thermionic valves. Transistors, integrated circuits, LSI, and microprocessors followed. Greatly to his credit Peter kept abreast of the rapidly changing technology. Importantly, he was able to impart this knowledge to students both locally and internationally.

Links with the profession and industry were a feature of Peter's work. He was chairman of the I.R.E.E Tasmanian Division in 1967 and 1968. He was a member of the Electrical College Board of the Institution of Engineers Australia and chairman of the Instrumentation and Control Committee.

Peter was head of Department of Electrical and Electronic Engineering for 6 years and engineering degree coordinator for 5 years. He established the Computer Systems Engineering BE course in 1985.

Peter's research interests included instrumentation, measurement, embedded microprocessor systems, borehole survey systems, seismic instrumentation, and condition monitoring. His capacity for the creation of electronic equipment led to significant consultancies outside the university. He installed the area's first volcanic eruption warning system in Rabaul, Papua New Guinea; Peter designed remote seismic recorders at the ANU in Canberra and deployed them across the Nullarbor together with low-power data acquisition systems and monitoring equipment. Peter, together with his colleague the late John Brodie, built equipment for underwater survey and ship movement monitoring following the Tasman Bridge Disaster. One of his instruments sits on the bow of the ill-fated Lake Illawarra at the bottom of the River Derwent; the other in the base of the Gordon Dam is known as the Watt inclinometer.

Peter's later focus was on bringing engineering education to parts of Asia. He consulted in Malaysia and in Vietnam where he was associated with the Engineering Twinning Program with the University of Technology, Ho Chi Minh City, and also at the National University of Vietnam. Recently, Peter delivered a second-year unit to engineering students at the Shandong University of Science and Technology in Jinan, China, in front of a class of 100 students for 36 hours each week for 2 weeks in the coldest part of the year with his wife's support—a Herculean effort. Students at the University of Tasmania and elsewhere whom Peter taught, and staff who worked with him, will remember him as a dedicated teacher and a very professional and capable engineer.

Peter Watt contributed the following account of his life as an engineering student in the 1950s:

I began as an engineering under-graduate in 1953 and continued for the next five years graduating BSc and BE(Hon). The comments I have written here are as I remember them, but with the passing of many years may not be absolutely accurate. The University campus and the engineering course was much different than it is today. In fact, life in general was much different. In those days, the University was essentially situated on two campuses, the Domain Campus and the Sandy Bay Campus. The Domain campus was the original University site and when the University outgrew the original buildings departments started to move to the new site situated on the old rifle range in the suburb of Sandy Bay. When I was a student, the Departments of Mathematics and Physics had moved into temporary accommodation on this site. The accommodation consisted of temporary pre-fabricated wooden huts housing Mathematics and a temporary lecture theatre and huts for Physics. We knew these huts as 'Nissen Huts', but they did not have a semi-circular cross section, the hall mark of the original Nissen Hut. Mathematics was about where the Staff Club is now and the Physics was situated somewhere near the top sports oval, towards the current Law building. The engineering course was 4 years with the first year a full science year plus two additional units, Engineering Drawing and Design and Engineering Workshop Practice. The second year was about equal parts science and engineering units. The third year was engineering course with specialization only in the fourth year. In first year, we had classes in Pure and Applied Mathematics, and Physics at the Sandy Bay site, Chemistry lectures at the Domain, and Chemistry laboratory, Engineering Drawing and Design and Workshop Practices at the old Hobart Technical College in Bathurst street adjacent to the University Domain site. It was necessary to travel between the two campuses and the University had its own bus service to transport students. This consisted of a small bus we nicknamed the 'meat wagon' as it looked like a truck with a square box built on the tray and painted brown. Entry and exit was via a number of steps let down at the back of the truck. Not very comfortable but no cost to students.

I spent three years travelling between the two campuses (first two years of engineering and the third year of science). The last two years of engineering was entirely at the Domain site and the old Hobart Technical College. The Hydraulics and Thermodynamics laboratory was also at the College as were lectures in Chemistry of Engineering Materials. Structures laboratory was in a temporary hut behind the main Domain administration and library building. The ties between the University Engineering and the College were very close and I believe the salaries of some of staff were sourced from both institutions. We had a three-year common course for all engineering disciplines (Civil, Mechanical and Electrical) which included large slabs of science units. In fact, first year engineering consisted of first year science (Pure Mathematics1, Applied Mathematics 1, Physics 1 and Chemistry 1) plus Engineering Drawing and Design and Workshop Practice. We started lectures in Drawing and Design sometime in January, about 3 weeks before the beginning of term1. This was full time, 5 days a week. Workshop Practice was held every Friday afternoon in the Hobart Technical College Workshops and finishing at 6 pm. The second-year course consisted of pure and applied mathematics and physics (science units) and mechanical engineering, materials and structures and design. I had timetable clashes because I was enrolled for a double degree in science and engineering, and I could not attend any structures tutorial. My first ever bending moment diagram was actually drawn in the exam, and it was a miracle I managed a distinction in the unit. Third year included mechanical engineering, hydraulics, materials, structures, surveying, design, electrical engineering and chemistry of engineering materials. We specialized only in the fourth year and even then, we all completed Business Practice for Engineers as a single group. This was possible when one realizes that a large content of the current engineering syllabus was not developed at that time. Semiconductors were only just being developed and there were only a handful of computers in the world. Software as we know had not been invented. Life was much simpler!

Examination in the 50's was also different to today. There was no examination at the end of the first or second terms, only once a year at the end of term 3. The philosophy behind supplementary examinations was also different. All supplementary examinations were held about 6 to 8 weeks after the annual exams. This allowed further time to study in order to gain a pass in the unit. The process of examination was not as transparent as it is today. Many units had 2 or 3 examination papers to cover the years' work. First year chemistry was for example split into inorganic, organic and physical chemistry. I found out the hard way that if you got 70%, 70% and 40% in the three sections you failed overall and had to take a supplementary in the section you failed. Laboratory work was counted, but I never found out how much. Several units I took had 4-hour exams. Most third-year mathematics units had 4-hour exams and I can remember at least one engineering fourth year unit with a 4 hour exam. The one redeeming feature was a cup of tea and biscuits provided at the half-way mark. All exams were held over a two-week (10 day) period. It was common to have 3 exams scheduled in consecutive time slots over 2 days. 4-hour exams and 3 exams in two days would cause a student riot today. At the beginning of fourth year engineering, I was able to enroll in the engineering honors program. This consisted of an additional 3 months' full-time work at the end of the fourth year examination period, lasting until the beginning of March of the following year. We had one week holiday between Christmas and New Year. The project part of the honors program was in the form of a mini research project. The School had just purchased its first transistors at this time (at £5 each) and my project was to investigate the possibility of using transistors in a chopper stabilized low frequency amplifier suitable for the amplification of signals from seismometers. This later led to a career in remote sensing, including seismic arrays, volcanic eruption warning systems, borehole surveying and underwater surveying.

Several incidents stick in my mind as an undergraduate student. We had a first-year pure mathematics lecture at 9 am, and woe betide any student who arrived after 9 as the lecturer (the Mathematics Professor no less) always arrived at precisely 9:00 and proceeded to lock the door of the 'Nissen' hut after him. A student who tried to climb in through a window at 9:05 was nearly barred for one year.

We had to write an essay entitled 'Force, Mass and Weight' for both Applied Mathematics 1 and Physics 1 at about the same time. Of course, we all wrote one essay and copied (by hand, there being no copy machines in those days). I was very upset when the two marks given for the same essay were 4/10 and 8/10 (a fail and HD). How can this be I wondered. In hindsight and with a lot more maturity I can understand it now, but at the time it was devastating. I was also troubled by the first year Physics laboratory. We had an experiment on simple harmonic motion which consisted of measuring the position of a slowly moving light beam on a graduated scale (it was a light beam reflected from a mirror placed on a stretched string which was kept oscillating). We all knew it should be a sinusoid but with the inaccuracies of measurement some points were slightly in error. However, to gain good marks for the practical we had to eliminate these errors by fudging the results. The students who did this got the highest marks. I'm not sure what this teaches students.

Several incidents in later years stand out. The first concerned the Hydraulics Laboratory which had a channel filled with water for pump and flow measurements. On arriving one afternoon and having spent the best part of an hour setting up an experiment involving the water channel it was discovered that a cat had been experimenting several days before we arrived and discovered that cats cannot swim (at least this one could not). This explained the funny smell in the laboratory and the class was abandoned. The second incident occurred when there was suddenly a mass exodus of students from the Thermodynamics Laboratory. We were setting up an experiment involving the measurement of the efficiency of an internal combustion engine when we watched the instructor pouring a measured amount of petrol from an open beaker into a tank feeding the engine. The mass exodus was caused by the observation that the instructor was smoking a cigarette with the glowing end about 4 inches from the open beaker of petrol. (His explanation later was that the temperature of the burning cigarette would not be above the flash point as long as you did not draw on the cigarette and there was no need to evacuate). I do not think the OH&S people would agree with him.

I will relate one more incident. This occurred in Chemistry of Engineering Materials lecture. One of my fellow students who became a bit bored with his lectures had the latest small radio in his pocket with a wire to an earphone in his ear. The lecturer after seeing this was concerned for the students hearing and inquired about his health. One of the other students let the cat out of the bag and the lecturer decided to get even. During the next lecture he walked up and down the desks and when behind the offending student produced a large pair of scissors and cut the offending wire to the earphone. No more was said but the student never listened to the radio again in class.

The engineering students organized and ran the Students Engineering Society. We had regular lunch-time meetings and organized engineering field trips. About once a month we screened a 16 mm film, sometimes a drama or comedy, but more often than not a film with some engineering background. One must remember that this was pre-TV and in many ways I think the social interaction was better in those days than it is now. The Society organized field trips every year to visit engineering sites in Tasmania. This included visiting Hydro dams and power stations, the Launceston Railway Workshops, the mines of the west coast, the cement factory at Railton, pulp and paper mills as well as local firms (such as Cadbury's and Electrolytic Zinc) with large engineering works. Often these trips lasted 2 of 3 days and always between terms 1 and 2 or 2 and 3. We sometimes used the University 'Meat Wagon' as transport, a real test of endurance against motion sickness especially on the west coast road. I can remember one trip to Queenstown to visit the mines in the area. We stayed two nights at the old Queenstown Hotel, and on this occasion we were accompanied by the senior lecturer in thermodynamics and his laboratory assistant. They were given rooms on opposite sides and at the end of a corridor. On the last night we tied a piece of thick string between the handles of these two doors so that when one tried to vacate his room it would rattle the other door. We can only imagine what happened as the string was broken when we made an inspection in the morning. One must remember the Queenstown Hotel was not exactly 5 star, with bathroom facilities along the corridor, not en-suites in the rooms. The second incident occurred when we were taking a walk after dinner. We came across a Morris Minor car (from memory green) parked in a narrow laneway. We started to debate whether the lane was wide enough to allow the car to be parked across the lane. An argument broke out and to test the visual perception of opposing sides we decided we had to experimentally test the hypotheses. Several of the students were football and rugby players, so we had little problem lifting the small car. It turned out that there was just enough lane width to turn the car at right angles with about 4 inches to spare. Having settled the argument, we could see no point in lifting the car again so left it across the laneway.

Another aspect of the Students Engineering Society was to organize social events. The social life at the University included many clubs and societies much as it does now. Many of these groups organized regular dances and a ball once a year. The Students Engineering Society was no different and we had one of the best balls of the year. I can remember on one occasion the students one year above me developed a sex detector for that year's ball. This consisted of two copper plates each one foot square and placed two feet apart on the floor. To detect the sex of a person you stood on these two plates. There was a box behind you and another one in front. On the front box there was a handle which you squeezed (it contained a pressure switch) and one of two lights lit up. Left for male and right for female. It worked every time and was 100% correct. One must remember that in those days men wore trousers and women wore skirts or dresses, especially at a ball. The detection system consisted of an infra-red beam about one foot above the floor between the two boxes. The beam was broken by a skirt but not by trousers. I think it was a mystery to some males and most if not all females at the ball.

There is a play called 'The Department' by David Williamson (himself an engineering graduate of Melbourne University and one time lecturer at Swinburne) that I saw in Hobart in the late 60's (or early 70's?). I am sure he drew most of his ideas for the play from his experiences at Melbourne and Swinburne Universities, however many of the staff and operations depicted in the play bear a remarkable resemblance to the Hobart scene. I would recommend this play to anyone interested in university life in general and in particular about engineering schools around the 60's.

Looking back on my undergraduate years, I am thankful I had so many learned lecturers and professors guiding and inspiring me. Many of the staff I can only describe as wonderful eccentric stores of knowledge, most of which have unfortunately passed on.

4.8 The "1980 Rule"

In the early 1970s, the Institution of Engineers, Australia, gave notice to all tertiary institutions that by 1980 only 4-year degree courses would be recognized as satisfying the requirements for full corporate membership of the Institution. The so-called 1980 rule triggered the establishment of engineering degree courses within the newly emerging Colleges of Advanced Education in Australia. Engineering degree courses were established within the Tasmanian College of Advanced Education on Mount Nelson, just up the hill from the university. The legacy of the Orr case (Eddy, 1961) saw several prominent Hobart engineers actively promoting the TCAE engineering course in direct competition to the university. The assertion, which may have had some merit, was the UTAS engineering course was too theoretical, and science based, whereas the TCAE course trained practical engineers who were immediately productive when they graduated.

The TAFE system was established with the Tasmanian Education Department with TAFE campuses teaching 2-year post-year 10 associate diploma courses in Hobart, Launceston, Devonport, Burnie, and Queenstown.

Matters came to a head (not just for engineering) when the university convinced the Karmel inquiry into tertiary education in Tasmania that the concentration of tertiary courses in Hobart should end (Australia, 1976). A new Tasmanian State Institute of Technology (TSIT) was established on the Newnham campus in the City of Launceston adjacent to the newly chartered Australian Maritime College (Fig. 4.4).

Professional engineering at the TSIT relied on joint teaching and resource sharing with the AMC. Students transferred after 2 years at the TSIT to the (then) Swinburne Institute of Technology which offered an attractive sandwich course with students spending 6 months in full-time industrial work as part of their degree.

Both the TSIT and the university admitted students with TAFE associate diplomas to their degree courses with somewhat less than a year's credit for their TAFE studies. The perception of the university staff was that the TSIT course was of somewhat less academic standard than the university course, with students admitted to the TSIT course who did not meet the university entrance prerequisites. On the other hand, there was a strongly held view of the part of the TSIT staff, and a significant number of professionals, that the university courses were too analytical and did not produce graduates with the technical engineering skills and knowledge sought by employers.

Fig. 4.4 Australian Maritime College, Newnham campus c. 2018

4.9 Engineering at the University Post-Karmel

By the mid-1980s, Professor Arch Oliver who had been appointed "professor of engineering other than electrical" in 1958 had recently retired. His replacement was a mechanical engineer with a strong research track record in fluid dynamics and acoustics. At that time, the "perfume" formula 4711 was used for the division of resources between the three departments in the faculty: surveying with four full-time academic staff, electrical engineering with seven, and civil and mechanical engineering with 11.

Leadership in civil engineering (as distinct from mechanical engineering) virtually disappeared with Professor Oliver's retirement. Oliver was originally a civil engineer, and the C&M Department during his stewardship had Malcolm Gregory as reader in civil engineering. When Gregory retired prematurely, his replacement was at lecturer level. The issue of academic leadership in civil engineering developed into a bitter issue within the C&M Department.

The retirement of the professor of electrical engineering, Professor Charles Miller in 1985, initiated a review of that department. The outcome was the establishment of a new department of electrical engineering and computer science. The review also recommended the establishment of a chair of electrical engineering.

The newly appointed professor of electrical engineering set about rebuilding and strengthening electrical engineering within the university. In a short time, he had sufficient support from within the faculty and the engineering profession to overturn the principal recommendations of the review that the professor of computer science should be the head of the new department for a period of 5 years. The result was the celebrated "divorce" between electrical engineering and computer science.

4.10 Engineering Education in Launceston

During the 1980s, professional engineering education was provided in Launceston by both the AMC (maritime engineering) and the Tasmanian State Institute of Technology which was running a 2-year feeder course to a degree at Swinburne University of Technology and sharing resources with the AMC located just 20 m away. As mentioned above, this situation was the result of the Karmel Report (Australia, 1976) a few years earlier which had closed down the Mt. Nelson campus of the Tasmanian College of Advanced Education in Hobart. The TSIT/Swinburne degree, having its roots in the CAE sector, was held by its proponents to be more practically oriented and relevant to industry's needs than what was, in their view, the overly academic and research-oriented university BEng. Research was not part of the TSIT profile; academic staff were encouraged to develop links with local industry and to be proficient and innovative teachers. The 2-year feeder course to Swinburne had strong appeal to students because of its 6-month industry placement sandwich structure.

4.11 Amalgamation and the New University

At the time the university and the TSIT amalgamated in 1990, there was residual antipathy resulting from the perceptions of each other's courses, but there was a commitment to make the amalgamation work. Engineering at Launceston was linked with architecture whose head, Professor John Webster, became a full professor within the amalgamated university. Amalgamation saw a complete revamp of the university's administration with the establishment of Vice-Chancellor Alan Gilbert's model of separate academic and administrative structures. Engineering (North and South), architecture (North), urban design (South), and surveying (South) were grouped within a newly formed School of Architecture and Engineering.

4.12 Course Changes in Hobart

Early in 1995 the vice-chancellor called a meeting of members of the School of Architecture and Engineering Committee for Quality Assurance to direct that the earlier specialization within the electrical degree stream, as requested by that department, should be implemented. The proposal was that students wishing to major in electrical engineering follow a separate second year course. This proposal was opposed by the Civil and Mechanical Engineering Department who saw the common first 2 years as one of the strengths of the Tasmanian BEng. There was also a substantial shift of load from the C&M Department to the Electrical and Electronics Engineering Department resulting from the changes to the second year of the course and a bid by the E&E Dept for a share in the teaching of management units. A rift developed between the two Hobart-based departments, and the relationship between the two heads of departments was at times acrimonious. At the time I was head of engineering in Launceston and for a short period head of architecture. My proposal to move civil engineering to Launceston to sit alongside architecture was, not unexpectedly, vetoed by the head of civil and mechanical engineering in Hobart.

4.13 The McKinnon Walker Review

In 1997, as part of the quality assurance procedures of the university, Emeritus Professor Ken McKinnon and Ms. Sue Walker conducted a review of engineering education in Tasmania (McKinnon Walker Review). This was a comprehensive review of all engineering-related courses in the state and the relationships between the major providers, namely, the university, TAFE, and the AMC.

The review recommended (in summary) with respect to cooperation and coordination:

- The establishment of an advisory Tasmanian Board of Engineering Education
- Close cooperation between the university, AMC, TAFE, and Tasmanian industry to ensure appropriate recognition of proper learning, credit arrangements, and articulation
- Continuation and extension of reciprocal exchanges for the use of laboratories and classrooms

With respect to the recommendations concerning the Bachelor of Engineering degree program, the school subsequently established an Engineering Curriculum Review Committee under the chairmanship of Professor Roger Tanner (the Tanner Committee). The terms of reference of this committee were to consider:

- The 4-year BE program, with a view to producing a common first and second year curriculum
- The BTech program
- Possible intercampus transfer and cooperation with AMC and TAFE

The Tanner Committee reported in April 1998. The outcome was a fully common first year, with 75% of the second year common to all disciplines. The remaining 25% of second year comprised a 12.5% discipline-related elective plus a 12.5% service unit taught in the second semester of second year.

Another significant outcome was the closure of my Department of Engineering in Launceston with the feeder program being run by the Australian Maritime College.

4.14 Level of Funding

The School of Engineering at the University of Tasmania has for many years been under strain due to the tightness of government funding available to the university and subsequently to the school. The present level of funding fails to provide adequate academic and technical staff levels for provision of proper teaching. As early as 2018, the dependence on full-fee paying international students was much too high and left the school with an overexposure in that area. Failure of the government to address the funding issues will result in a reduced capacity for the school and university to meet the need for engineering graduates. This problem is common to all Australasian universities and had a major impact on student enrolments and funding when China halted new students coming to Australia during the COVID-19 pandemic.

University funding for academic activities is predominantly directed toward salaries. Approximately 80% of recurrent expenditure is directed toward staff salaries. Thus, a good measure of relative funding for university degree courses is the ratio of the numbers of students to the numbers of academic staff in a particular degree course. In the period from 1997 to 2001, the ratio of equivalent full-time students (EFTS) to equivalent academic staff (EAS) in the School of Engineering blew out from 11:1 to over 16:1. From 2008 to 2011, the student staff ratio rose further to 21:1. This was largely due to the university policy of encouraging early retirements to ease the budgetary problems. This policy has an additional adverse effect in that senior experienced staff are being replaced by junior staff with little teaching experience. A disturbing feature is the growing reliance of the school on income from international students. In 1994 the ratio of international students to Australian Government-funded students was 26%. By 2000 this ratio had risen to 35%. This makes the engineering school very vulnerable to a downturn in the numbers of international students. Presently, the student staff ratio has improved as shown in Table 4.1 largely as a result of the reduction in international enrolments due to COVID.

Table 4.1 Ratio of equivalent full-time students (EFTS) to equivalent academic staff (EAS) 2018–2023

2018	2019	2020	2021	2022	2023 (forecast)
21.4	20.6	17.9	15.9	14.3	13.1

Source: School of Engineering. Application for Engineers Australia Accreditation 2023

4.15 Life As Engineering Students in the 1970s

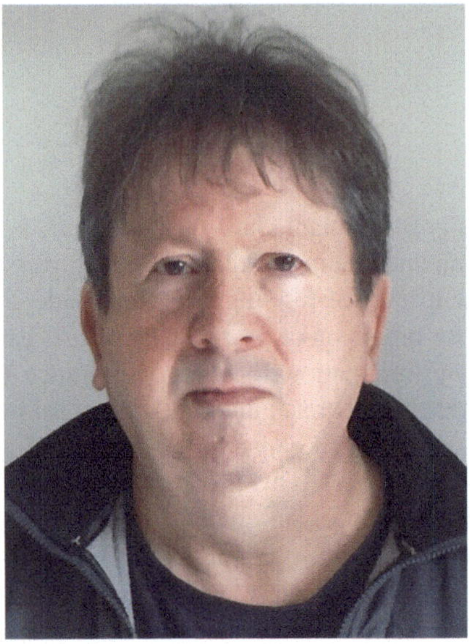

John Minchin (1976)

I knew a bit about the engineering faculty long before I started there in February 1971. My father grew up a few houses along Grosvenor Street and told me many tales of his exploits in the Remount Depot and the shooting range further up the hill. Then as a youngster when visiting my grandparents, I saw the construction of the engineering building progress. I knew I was destined to study there from the time my engineer uncle gave me a Meccano set for Christmas.

I remember we started early in order to do a 2-week technical drawing class before the official start in March. There I met Ted Sturges—he imparted lots of practical knowledge, and when you heard whistling in the corridor, you knew he was coming. Strangely, many students objected to having to learn to draw, whereas I had assumed you would need to draw what you were trying to design.

Most of the first year subjects were mathematics and science, but Engineering 1 was my favorite—Peter Doe was our lecturer, and we learned about gears, rotating machinery, and other interesting stuff. I was in Dr. Doe's Materials and Structures IIA class a few years later where I scored 100% in the exam. This was a rare occurrence in general but even rarer for me. I managed this around the same time as I was struggling with electrical engineering with Prof. Miller (I disliked it immensely) and fairing even worse with pure mathematics.

Professor Archie Oliver was very memorable of course. In a letter I wrote to him 20 years ago, I recalled how I was amazed at the way he could fill blackboard after blackboard with notes, equations, and analyses without reference to prepared notes, even more so when he was temporarily taking a class for an absent lecturer. He obviously loved what he was doing, both the teaching and the research, and this came across to the students and motivated us to want to learn. One of his frequently used sayings stuck in my mind: "There's no substitute for capacity." I have occasionally borrowed and used this phrase myself over the years.

Greg Walker had a quiet demeanor but opened up when you showed interest in his work. Malcolm Gregory was a very clever and inspiring man. We loved many of his anecdotes (one I recall was about how the French rail gauge came about in order to accommodate the president (?) needing to sleep crossways in his carriage) and the interesting annotations and comments on our assignments and lab reports. One of the best comments I received was "useful work." I met him in the street 25 years later—he remembered me and was genuinely pleased to hear what I had been doing since graduation.

Then there was Eric "Eccles" Middleton, a great friend of students and very at home with a glass in his hand. While my brain forgets names from last week, I can still remember Dr. Rish (who taught us surveying), Alan Christian who gave us practical skills in the workshop, and Godfrey Burrell in the thermodynamics laboratory. After he retired, Alex Pittas arrived and tried to teach us mechanical engineering. I was impressed with his pathway from being a migrant working as a cleaner to an engineering professor. Our paths crossed many times over the years—he was a real gentleman.

Others I recall include Colin Foster and Frank van der Woude. I don't remember the electrical engineering staff as clearly, mainly, I think, because I never felt comfortable with electrical engineering, particularly in the practical sessions.

During my time studying hydraulics, I spent time carrying out practical work in the hydraulics laboratory, which was a partnership between the engineering school and the Hydro-Electric Commission (HEC). Here I met Frank Navin, and some years later when working in the Hydraulics Department at the HEC, I had firsthand experience of hydraulic modeling when working on the design of the Murchison Dam spillway.

Early in my studies, we received a fascinating lecture from Dr. Sergio Giudici from the Hydro-Electric Commission about the unique design of the Gordon Dam. Years later, I was fortunate to work closely with him during the formative years of the HEC Consulting Business, later to become ENTURA. I learned much from him about people management in a technical environment. I also learned about amateur radio, salami making, espresso coffee making, and many other topics where he had knowledge and expertise.

I was successful in obtaining a Hydro-Electric Commission University Scholarship for my engineering studies and, along with Michael Green, became the final recipient of that award. A benefit (along with the stipend) was guaranteed vacation employment during the course. This employment was a valuable element of the course—learning real engineering skills but more importantly learning to work with other people: engineers and technical and trades people. The money came in handy too!

I enjoyed my time studying engineering at UTAS. At times it was a struggle, especially with pure mathematics, but I always felt that the engineering faculty staff wanted us to succeed and become productive members of the engineering world.

In May 1976, I was offered a position in the Hydraulics Section of the Hydro-Electric Commission. This was rather strange as I had studied mechanical engineering and this was a "civil" role. But I got by. The most interesting work I undertook was hydraulic design and layouts for Murchison Dam spillway, Bastyan Dam spillway, Reece Dam spillway, and Sophia Tunnel.

In 1978 I also became a part-time lecturer in applied mechanics at Hobart Technical College.

I was asked to spend a couple of years as a construction engineer on the Pieman River Power Development. In early 1979, following my marriage, I moved to Tullah where I worked as a site engineer at Murchison Dam supervising construction of access and haul roads, dam plinth, and the Sophia Tunnel intake gate structure. Interestingly, I supervised excavation for the spillway I had helped design. My wife taught at the local primary school.

In 1980 my mechanical credentials were uncovered, and I transferred into the plant section. Firstly, as permanent installation engineer, I carried out planning, assembly, installation, and testing of the permanent mechanical equipment including gate frames, hydraulic wheeled gates, bulkhead and radial gates, and trash racks. A standout highlight of this work was planning and supervising of the closure of the diversion tunnels at three dams.

Over the next few years, I was responsible for field maintenance crews (hydraulic and pneumatic drilling rigs and tunneling equipment, hydraulic winches, dam slip forming equipment, and tower cranes) and the main repair workshop maintaining major earthmoving equipment.

In 1987 I transferred to Queenstown as field planning engineer for the King and Anthony Power Schemes (including construction support services, maintenance and operation of village and camps, and construction restoration and revegetation). A significant function was the implementation of computerized work scheduling and costing throughout the construction works.

In 1991 I transferred back to head office as a senior Gates design engineer. The highlight of this time was liaison between design, workshop, and construction personnel for the manufacture, transport, and installation of the Crotty Dam dewatering outlet ring follower valve and jet flow gate. This included design and planning for site installation equipment and on-site technical advice and supervision during a cold and snowy winter.

During this period, I had commenced an MBA by distance learning. In 1992 I managed the Design Support Department covering a wide range of support functions including estimating, job costing and financial systems, budgeting, contracts and procurement, drafting, and quality management systems. A highlight of this time was the commercialization project, and I was lucky to be involved in a number of projects including business reorganization, implementation of quality management, digitization of engineering drawings, and the planning for the merger of Consulting Business Unit and the HEC Enterprises Corporation.

In 1995, as manager of the Business and Project Services Group providing a range of support services to the Consulting Business Unit, I was fortunate to work closely with the inaugural general manager, Dr. Sergio Giudici.

Following a reorganization in 1999, I took on the role of project manager for the project to select and transition to a single outsourced IT service provider for Hydro Tasmania, Aurora Energy, and Transend Networks. I learned much about how to manage a project, particularly the importance of stakeholder communication and management.

During the period from 2001 to 2011, I worked as Manager (Business Risk and Policies) and Business Risk and Compliance Manager. The main part of these roles was developing and maintaining the organization's integrated risk management policies, programs, and systems and organization-wide insurance. In addition, I managed the integration of compliance policies, programs, and systems with the risk system.

A challenging but interesting part of the work occurred during the period following the September 2001 attacks and later in 2009 when the H1N1 pandemic occurred. Using a risk-based approach to convince global insurers of our ability to reduce and control our risks enabled escalating premiums to be controlled. At the same time, I took a lead role in the development and management of emergency, business continuity, and security management systems in cooperation and collaboration with local and interstate electricity entities, SES, and Tasmania Police.

Following my 35-year career with Hydro Tasmania, and a further year working on contract with Transend Networks, I determined that any postretirement work would be geared toward giving back to the community. Since 2012, these roles have been the following: member of the Board of Hobart City Mission since 2013 and president since 2018; member of the Board of Hobart City Mission Foundation from 2020 to the present; president, Tasmanian Association of State Superannuants Inc. from 2019 to 2023; and member, Tasmania Branch Committee, Association of Professional Engineers.

Michael Green (1973)

Michael Green was an undergraduate from 1970 to 1973 graduating as an electrical power engineer. This was followed by research in the application of superconductivity in direct current machines and finally receiving a Master of Engineering Science in 1982. With Jack Beresford's supervision and assistance from the technical staff Bob Barclay, Steve Avery, Shaun Cruikshank, and Garry Owens, we were the first to achieve superconductivity in Australia for such an application. Air-freighting liquid helium from the mainland presented a particular logistic challenge to make sure most had not evaporated before it arrived.

Attending UTAS was a bit like an extended family with small classes and ready access to lecturers. I look back and think of those with the most influence on my studies. In the forefront was Jack Beresford who conveyed concepts clearly and I believe was the savior of many students. The uncle figure was Ted Sturges in the drawing office. He was our first contact in the engineering faculty as first year studies were mainly in the science faculty.

Other engineering lecturers who influenced me were Dr. Rafael Rich who inspired me to almost study structural engineering, Prof. Miller and his almost vintage Jaguar, Bob Wherret, John Norton, Peter Watt, Eric Middleton, David Lewis, John Brodie, and Godfrey Burrell with his old steam engine in the thermodynamics lab. Professor Archie Oliver with his old coat and pipe and beloved holden car taught us to think for ourselves. Any questions in the lab or during tutorials with Dr. Walker would be met with general direction as to the answer and left us to discovery for ourselves.

Other memories that compare to where we are today with information technology are the arrival of the PDP-8 computer where the big advancement was moving from a "ticker-tape"-type reader to a high-speed reader saving half an hour loading the operating system. The other was the Olivetti programmable calculator in the structures lab that relied on "splitting" memory locations to provide space for code.

There was enthusiastic competition between the law students and engineers as we shared the building, in particular the main lecture theater. Consequently, some lectures were held in the makeshift, metal clad Musset huts before the law buildings were constructed. The hot summer conditions in the Musset huts were challenging, in particular sitting for tests. The annual engineer's challenge chariot race from the campus to Franklin Square heightened the competition across faculties, in particular with geology and law.

In summary, studying in the UTAS engineering faculty was a very personalized experience, and we were privileged to be so well treated. The Colombo Plan made for a very diverse faculty with students from across the Southeast Asian region and also provided us with diverse cultural experiences.

Professional Affiliations: Fellow, Institution of Engineers Australia; chartered professional engineer CPEng; registered professional engineer, NER APEC Engineer IntPE (Aus); engineering executive EngExec; past president, Tasmanian Division I.E. Australia; past Tasmanian National Congress representative, I.E. Australia; past member of CIGRE Australian Panel APC5 and predecessor APC1, 2000 to 2009; and past member of CIGRE International Working Group WGC6–2.

Experience as an Independent Consultant

Independent consultant since 2009 providing services in the areas of Australian national electricity market rules and legislation, effecting generator and load connections to transmission and distribution systems, power system planning, electrical power engineering, plant commissioning and testing, commercial negotiations, and regulatory registration and compliance.

Experience with Transend Networks Pty Ltd

After Tasmania's successful entry into the Australian National Electricity Market in May 2005 and commissioning of Basslink in April 2006, I continued as a member of the executive management team with responsibility for project closure issues and manager of Business Planning Regulation and Compliance (later executive manager of Corporate Strategy and Compliance) responsible for executive management of Transend's interests in economic regulation and compliance with regulatory instruments.

I was executive manager of Tasmanian Wholesale Electricity Market and member of the executive management team with responsibility for the successful integration both technically and commercially of the interconnection between Tasmania and Victoria, Basslink, into the Tasmanian power system and providing the interface with Basslink Pty Ltd, the proponent of Basslink.

I was manager of Development and Regulation and member of the executive management team responsible for the development of detailed expansion/refurbishment/replacement plans for the Tasmanian transmission network, interface with state and national regulatory bodies in relation to revenue and performance determination and monitoring, and commercial and technical negotiations with transmission network customers.

Experience in the UK

Experience in the England and Wales competitive electricity supply industry.

Experience with Hydro-Electric Corporation

Senior analyst providing advanced technical knowledge and survey and analysis of external environments including socioeconomic, political, technical, business, and markets.

As part of national reforms of the electricity supply industry, convened a national transmission pricing working group with responsibilities for the development of principles for pricing of and access by participants in a national electricity market to the natural monopoly transmission and distribution systems.

Electrical investigations engineer responsible for the management of a unit with the functions of providing studies of alternative power and system developments, determinations of quality and security of supply, recommendations on system stability, governing, AVR characteristics, harmonic analysis, insulation coordination, load flows, and fault analysis.

Project engineer responsible for the electrical design and control (including AVR and governor) aspects of King Power Station (Tribute Power Station) including power station physical layouts, 220 kV transformer arrangement, design of power station major systems, technical administration of generating plant contract, and coordination of department work and all commission work associated with layout and electrical aspects.

Responsible as site representative for the commissioning of electrical design aspects of Pieman No. 2 generating plant.

Responsible for installation and operation of 22 kV lines, 22 kV/415 V pole and ground-mounted substations, diesel generators, and 415 V reticulation for a large civil construction site.

Construction site responsibility for the installation of 110 kV and 220 kV switchyard equipment and associated control and protection equipment.

Responsible for performance evaluation under fault conditions of system transmission, transformation and generating plant protection, preparation and implementation of protection settings, specification preparation and tender evaluation of transformer and bus bar protection, and site commissioning of generating plant protection.

4.16 Relocating the Sandy Bay Campus

In the first decade of the twenty-first century, the University of Tasmania embarked on a policy of relocating its activities to the Hobart Central Business District (CBD) and selling off the Sandy Bay campus for a major real estate development.

Hobart is a comparatively small city and relatively flat away from the Wellington/kunanyi and Mt. Nelson foothills. The proposed city-centric plan was to relocate staff and students over an area in the Hobart CBD along a corridor of Melville Street to the original buildings on the Domain. The Medical School, the Conservatorium of Music, the School of Art, and the Institute of Marine and Antarctic Studies (IMAS) were already close by on the waterfront.

These buildings formed a triangle of side length roughly 1.4 km. The physical sciences and engineering buildings on the Sandy Bay campus are only 2 km from IMAS building by the most direct route and about 3 km from all other university-owned building in the city, about 15 min by bicycle or scooter. Keeping the physical sciences and engineering buildings at their present location on the Sandy Bay campus would extend the university's footprint in the city by at most a couple of kilometers and include the prime residential, shopping, and tourist precincts, Sandy Bay and Battery Point.

4.17 Reimagining the Sandy Bay Campus

In a submission to an invitation by the university to "reimagine the Sandy Bay campus" in August 2020, I wrote:

> The campus below Churchill Avenue area might well be described as the heart and soul of the Sandy Bay Campus. The buildings are pleasingly gathered around the grassed central spine which is adorned with art and artifacts of significance, including the Les Baraclough fountain and the Tony Sprent sundial. Viewed from the river, the Campus rises from the green apron of its sports fields through the impressive Centennial Building and is crowned by the panorama of the Humanities, the Stanley Burbury Centre, Administration Building and the Morris Miller Library. These buildings and their co-location have aesthetic merit and historical significance for generations of UTAS alumni. The Campus below Churchill Avenue must be preserved.
>
> The Sandy Bay Campus of the University of Tasmania should be redeveloped and repurposed into an attractive learning, research, commercial and residential precinct including the present Physical Sciences buildings, updated with modern equipment. (Doe, 2020)

4.18 The Future of STEM

An elector poll of ratepayers in the city of Hobart in October 2022 resulted in a 74% vote of no to the question: "Do you support the University of Tasmania's proposal to relocate the Sandy Bay campus into Hobart's central business district?" An effective and concerted campaign by the "Save the UTAS Campus" group resulted in Tasmania's government promising to introduce a bill requiring a majority of both houses of parliament to agree to the sale of any part of the existing Sandy Bay campus. The university will survey staff and students on the location of a proposed STEM (science, technology, engineering, and mathematics) facility. It now confirmed the existing physics, geology, chemistry, and engineering buildings on the Sandy Bay campus will be refurbished and re-equipped to be the core of the UTAS STEM teaching and research facility (Rufus Black, vice-chancellor, pers.com. 28 May 2024).

4.19 University of Tasmania: A Special University

What is it that makes engineering at the University of Tasmania special? What is it about this place that fostered my rewarding career and the careers of many exceptional engineers and leaders?

It can be attributed to the quality and dedication of the lecturing staff, to the relatively small size of the school, and to a commitment to breadth and hands-on training. The BEng degree in my time as a student had 3 years of common course, with specialization in the fourth year. This has been reduced, over the years, to one common year, but the University of Tasmania BEng still has more breadth than most Australian and international engineering degrees. The vision of the early professors of engineering, Burns, Oliver, Newstead, and Miller, ensured that the laboratories in the new engineering building on the Sandy Bay campus were spacious and well equipped. Civil engineering students were able to design, mix concrete, and make beams that were tested to destruction. This followed on from the pioneering work of Sir Alan Knight, a University of Tasmania engineering graduate, who, as well as designing Hobart's floating bridge, invented the composite steel and concrete beam and was later the chief engineer of the Hydro-Electric Commission. The "Hydro" was a major employer of engineering students and graduates. It sponsored the University of Tasmania Power Laboratory and a high-voltage test facility.

The engineering school has had its share of Rhodes Scholars including Dr. Sergio Giudici who was responsible for the design of the Gordon Dam and Dr. Stephen Gumley who was the first chief executive officer of the Australian Defence Materiel Organisation. Dr. Gumley provided the following synopsis of his illustrious career and a reflection on his time as a UTAS undergraduate.

Stephen Gumley (1979). Dr. Stephen Gumley AO, HonDEng(Tas) D.Phil(Oxon) MBA(Tas), FIE(Aust), FTSE (1979)

Following being dux of Tasmania in the 1974 grade 12 matriculation examinations, Dr. Gumley graduated with a Bachelor of Engineering with first class honors from the University of Tasmania in 1982. He then completed a Doctor of Philosophy degree in 1982 at St. Catherine's College, Oxford, as Rhodes Scholar. He returned to Tasmania to become a founding director of the internationally successful CAD company, Protel Technology (now renamed Altium). His career in the late twentieth century included time working with the technology and manufacturing companies Critec Pty Ltd and as CEO of the Global Lightning Technology Group in Hobart.

He moved to Melbourne as chief executive officer with the Preston Group in Melbourne, an international exporter of advanced Australian developed simulation and modeling information technology products for the aviation industry covering airspace simulation optimization and airport operations. He continued with the aviation industry with Boeing in Seattle where he was vice president (Information Services) responsible for a new division with US$1.8 billion in recently acquired assets including integration of recent mergers and acquisitions.

In July 2002, Dr. Gumley was appointed chief executive officer and managing director of the Australian Submarine Corporation (ASC) with responsibility to transition ASC to become a whole-of-life logistical support organization including modifications, R&D, maintenance, capability enhancements, and training. ASC provided Australia with six fully operational Collins-class submarines. He was CEO of ASC during the building and outfitting of the sixth and last Collins-class submarine HMAS Rankin.

His military responsibilities expanded in February 2004 when he became the chief executive officer of the Defence Materiel Organisation (DMO), Canberra, which acquired and sustained nearly all of Australia's specialized military equipment. While Dr. Gumley was CEO of DMO, it had 6000 civilian and 1500 military staff, around $120 billion of business under management, and an annual cash budget of over $11 billion.

He was honored as an Officer of the Order of Australia (AO) on the 26th of January 2010. The citation reads: "For service to public sector management, particularly through leadership of the Defence Materiel Organisation, and the development and implementation of significant reforms in procurement and sustainment of military equipment. On 7 July 2011, the Australian Parliament accepted a Statement of Indulgence from the Minister for Defence, supported by a Shadow Minister, about Dr. Gumley's contribution as Chief Executive Officer of the Defence Materiel Organisation" (House of Representatives Hansard pages 8–10, 7 July 2011, pdf pages 24–26).

Dr. Gumley has successfully performed a variety of chief executive roles since leaving the DMO with experience in both the private and public sectors.

He was CEO of Hutchison Ports Australia from April 2012 to September 2014 responsible for construction of the third container terminals at each of Port Botany Sydney and Brisbane ports and the initial works on an intermodal train/truck freight facility at Enfield in Western Sydney. These $700 m + of greenfields infrastructure projects involved planning/environment, tendering, program management, stakeholder engagement, engagement with regulators (competition and security), and systems integration. Brisbane and Sydney container terminals were progressively opened for Australian trade business with large container ships from August 2013 and April 2014, respectively.

From October 2014 to May 2015, he was interim CEO of Murray Irrigation Limited, Australia's largest private sector irrigator, responsible for over 2500 km of infrastructure channels, gates, and control systems, irrigating about three quarters of a million hectares of prime farmland with 700 GL of water in southern NSW to about 1200 landowners. In addition to repairing its $200 m Commonwealth-funded infrastructure upgrade program, especially program management, Dr. Gumley conducted an extensive health check on the company while it sought a permanent CEO and initiated significant changes in management approach to handle a changed strategic environment. In July 2017, he was invited back to consult as an advisor to the CEO and board committees on the infrastructure projects.

From October 2014 to December 2016 and from October 2017 to July 2018, he was a member of Defence Council Victoria, an advisory committee to the State Minister for Economic Development.

From March 2015 to 2020, he was appointed a nonexecutive director of AMOG Consulting, a Melbourne-based professional engineering firm specializing in hydrodynamics, safety engineering, and marine structures.

Between January 2017 and June 2017, he was interim CEO of ANZ Terminals Pty Ltd, the leading independent fuels and chemicals port storage firm with facilities at Melbourne, Port Botany (Sydney), Geelong, Adelaide, and four sites in New Zealand.

In March 2018, Dr. Gumley commenced a 3-year term as a nonexecutive director of Tasmanian Water & Sewerage Corporation Pty Ltd (TasWater) where he was also a member of the Audit and Risk Committee and Capital Works Committee. He is presently chairman of TasWater—a 6-year appointment until 2024.

Dr. Gumley has reformed organizations, organically grown businesses, delivered major projects, turned around failing existing businesses, and was a public company director during a successful initial public offering (IPO). His skills are based around leadership of complex organizations, the public/private sector interface, strategy formulation and communication, industry policy, mergers and acquisitions (M&A) activity (Australia and overseas), professional services, corporate governance, industry regulation, strong commercial acumen, major procurement, contract creation and management, innovation and structuring, program/project delivery, financial judgment, international linkages, and keen understanding of people and organizational motivation.

As a comment on his time as a UTAS undergraduate, Dr. Gumley wrote: I am extremely grateful for the professional development given to me by Prof. Arch Oliver, Dr. Greg Walker, and Dr. Peter Doe. During the 3 months of Eng Honors (which happened at the end of fourth year engineering course in Dec and Jan/Feb 1979 before going to Oxford at the end of Sep 1979), they taught me research approaches and program management techniques. They went the extra mile and taught me how to undertake lab-based research efficiently and effectively. This was a huge benefit at Oxford where working in big wind tunnels I managed to hand in a DPhil thesis in 23 months after leaving Australia (skipping master's degree). I noted that in those years before personal computers, the Internet, and emails, with all papers stored in physical libraries that needed to be attended almost daily, most DPhil candidates took 3 or even 4 years from their bachelor degrees. I have always thanked the University of Tasmania's engineering faculty (including the helpful lab technicians). The faculty and its people were really good at the end of the 1970s decade.

Dr. Jane Sargison's doctorate from Oxford University on the cooling of jet engine blades built on Dr. Greg Walker, Prof. Oliver, and others' long association with Rolls-Royce and their research on the large (2 m diameter) axial compressor in the school's aerodynamics laboratory. Jane was named National Professional Engineer of the Year in 2011. Her address to the Tasmanian Division of the Institution of Engineers, Australia, on the occasion of the 2012 Alan Burn Memorial Lecture touches on the role of women in engineering and her career as a successful engineer and leader.

Jane Sargison (1988)

"What do Rolls Royce jet engines, the Rainbow Bee Eater Project, Education, and Tasmania have in common?" The inspiring story of a Tasmanian business and community leader. Good evening and welcome engineers, friends of engineers and interested people. As I look around the room here tonight I see faces of friends, colleagues, students and mentors. Many of you have long ago blurred the lines between those groups. And I thank you now for the role you have played. Tonight I will share with you reflections on my career so far. It has been a great journey and Engineering week is an appropriate time to reflect on the challenges and opportunities ahead for our profession. Big on my list of career highlights was hearing my name announced as Engineers Australia Professional Engineer of the Year, last year. It was unexpected. On the night I had barely prepared an acceptance speech given the exceptional field of finalists that I was judged against. The most significant aspect of this award to me was that it was an award for the whole field of professional engineers, not any particular category and not female engineer. Now don't get me wrong, female engineers appreciate, respect and are grateful for the efforts of many who have put together scholarships, prizes and other works to try to encourage women into engineering. But as our national president David Hood recently put it, this just might be like pushing our product (engineering) ever harder at a target audience (women) without really looking closely at why they haven't been taking up the product. The other downside to pushing the product is that the women who do participate in engineering could feel marginalized.

My first experience of the delicate balance that needs to be made in promoting and encouraging women into engineering was day one of University on walking into the engineering School. The School had also just started scholarships for women and had responded to the need to provide female toilets in the building by developing a large female only common room attached to the toilets, complete with small kitchen, comfortable chairs and heating. Now this was a delicate issue for the female students. The work that was done in providing scholarships and the facilities, addressed a significant issue—the school had to provide facilities and address the gender imbalance. But, the extent of the facilities which allowed the female students to enjoy a hot lunch in a warm, comfortable common room raised the ire of the male students. It didn't really help us to just fit in and get on with the job. The message to me was and remains that we need to be extremely sensitive in how we address the gender imbalance. This is again why the engineer of the year award was so exciting for me, because the issue of gender was removed from the process.

The problem of how to address the gender imbalance in the profession has captured the efforts of many. I am actually not the best person to give the answer to the problem, because I am an anomaly. I'm a left-handed female engineer. That puts me in the tail of any normal distribution, and for the non mathematical in the audience—I haven't come across many other female engineers in my career and hit more 'firsts' than I had expected. Incidentally I have come across more left-handed engineers than female ones. But having made it here as an outlier, let me reflect on some of my observations and experiences. Being an engineer is a lot about having a very particular brain. As one senior academic I sat next to at dinner recently put it, if an engineer says that 63% of Hobart's water came from the mountain catchment this month, then it is not 64% or about 60% but definitely 63%. Apparently there is prevalence of aspergers syndrome amongst the engineering population, this gives some indication of the end of the bell curve that engineers inhabit.

This mindset in the engineer is identified in mathematical ability and attention to detail and yet in practice it is closely linked with musical, artistic and creative talents. Ladislao Reti, the editor of the Codices of Leonardo da Vinci said of his work that he learnt:, "da Vinci was an engineer who occasionally painted a picture when he was broke" (Gies and Gies, 1994). The point of this particular brain is two fold: 1. it is particularly evident amongst males - be it culture or evolution we may never know, 2. it perpetuates many of the negative caricatures engineers are known for - lack of communication skills, lack of diversity, too much focus on one solution, no room for external influences in design. Just the sorts of perceptions that tend to turn your average teenaged female off the profession. And not an accurate perception of engineers and engineering. Perception is powerful. To give you an example, although I was enthralled by thermodynamics, energy and power, throughout my engineering degree I always felt that I had to work hard at the academic side, to make up for the fact that I wasn't overly interested in the idea of completing an engine rebuild of my car. It was not just my misconception that engineering students should have looked deeply into the workings of their vehicle, it was stated by more than one lecturer that there was a time when all engineering students would have taken their engine apart. I toyed with the idea of this, and I did in fact assist with minor servicing of my pride and joy—a red 1976 Toyota corolla coupe, but this was not my passion and in my busy life I just never seemed to find time to learn these skills.

So, I worked away at the academic side of engineering—and gradually, I realized that (actually) most of the boys in my classes had never stripped down an engine either. Given the complexity of the modern car, this trend will continue. The glorious days past of students working on their cars represents more about the cars, culture and lower wealth of the era, than the demise of the quality of the engineering undergraduate. My point is, there remains a misconception about what engineering is, and the characteristics you need to be a good engineer.

As I said perception is powerful and female engineers are influenced by it. Another observation that has become clear, particularly since I left academia is that not only are there lots of males in the engineering profession, but this is also true of the trades and professions that contribute to engineering activities. Reflecting on the disparity in female uptake of professions such as dentistry, medicine and law compared with engineering, it is apparent that medicos and lawyers spend their days at work with a greater gender balance than does the mine manager, the power station engineer and the chief of aircraft maintenance. Working in a university I could step outside the engineering building and experience gender balance in the campus café and clubs. On a mine site it is not so easy. So, we don't just have gender imbalance in the profession, but in our clients, our contractors and service providers. There have been numerous studies of the factors which attract women to study engineering, and the factors which lead them to stay in the profession. The single biggest factor: is having other women in the workplace. I was the first female mechanical engineer to be employed in the consulting firm where I took my first graduate role. I was the first female academic staff member in the engineering school at UTAS. Over time, however both organizations built up a small but effective cohort of female engineers. Of course I wasn't the only factor, but take it from me, if you want to attract women to your organization, make sure that you look after those that you have: they are your best advertisement. I look forward to the day when questions about female representation at high levels of management, boards and throughout the engineering industries are moot. But why wring our hands over having female engineers? Are they too much trouble? Is there ever going to be any greater participation than the few 'anomalies' we have at present? Are women just not interested in engineering? Diversity in the workplace isn't just about fairness, or political correctness. Diversity of gender, background, race and all the other range of variables brings with it new ideas and new solutions.

Early in the industrial revolution the engineer was revered in society and indeed engineers transformed society and the world as we know it rapidly bringing fresh water and sanitation, electricity, communications, transport and industry as we know it. However, increasingly engineering activities brought changes to the atmosphere, waterways and wilderness. The world has a mixed perception of the engineer now—the creator of the iPad and perpetrator of environmental damage. The world now faces an array of different challenges—how do we feed, house and provide power to the increasing population? The solutions we have used before are not going to be the solutions that we need now. How do we develop new solutions? As Albert Einstein put it—we can't use the thinking that created the problems to solve the problems. We need diversity. Diversity brings different perspectives. Diversity is about different characteristics—gender, race, background, but it is more than that. It's not about ticking a set of boxes as you look around the boardroom table. The other aspect of diversity I'd like you to consider is diversity of experience in the one individual. For example, I have worked in research, academia, international consulting firm and am now expanding that background working in small business, technology start up and government and company boards. Each of these roles requires me to understand and work within different constraints and with different objectives, to understand the needs, demands and ways of the different sectors. I think the ability to make these transitions with engineering training is something that we should both publicize and also encourage amongst our students, staff and colleagues. Moving through these roles also expands the engineers influence beyond the walls of the engineering design office and helps to spread the message about what engineers do and break down the stereotype.

It is also important to build diversity in experience across different fields of the one sector. Research is one such sector that gains enormously from looking outside the silos, although the structure of the workplace, the buildings and professional rewards are not yet up to do this. Consider this, engineering research—the very top cutting edge is extremely sophisticated. It's competitive, it's expensive. Often requiring significant investment in the equipment required to conduct the research. Large teams working for years to achieve the next small step of progress. This type of research is important. This research delivers the next 1% of solar panel efficiency, the next decimal place of control on medical equipment. But it is expensive, and the pathway from university research out into commercialization is not straight forward. How and when will the world get to use your work?

Training researchers to diverse multidisciplinary research is training Australians to one of our strengths. Collaboration and cross fertilization of ideas is incredibly powerful and there are many examples. This is something that I advocated when I was involved in university research. Particularly in Tasmania we have the opportunity through our smaller size, to make productive links between industries. For example: Engineering researchers working with architectural researchers—to improve the energy efficiency of multi-story office blocks and domestic houses. The conventional office block is designed by the architect with gaps left on the upper floor for air-conditioning plant to be installed, as an add on late in the project. Engineers can design more efficient aircon, lighting and acoustics within the box assigned but if they have the whole building then they can make step changes in performance. The deepest change occurs when the two professions work closely together. For the domestic house, there is research on the heat and mass transfers occurring in the sub-floor space. What impact does the ground, ventilation air and the many different materials and flow paths have on the temperature in a room. This research topic has captured architects, building scientists and engineers. The modelling and measurement of these transfers is complex and the equipment is expensive to develop. The returns on investment are not available in the housing industry. However, a small increase in efficiency less than 0.1% is significant in the aerospace industry and gas turbine research has poured enormous funding into accurate measurement and modelling of these heat transfer components. Taking the outputs from these great investments and applying to different areas yields great benefits.

There are many other projects that I have been involved with—engineers and agricultural researchers working on biomass energy and biochar project integrating renewable energy production with farming, forestry and carbon sequestration. Or the introduction of vortex generators used on every modern wing, into naval architecture. We cannot sit within silos. With diversity of technology, industry and culture we engineers can be more creative, more successful and more influential. The renewable energy sector relies very heavily on diversity of expertise and is one of my career foci. The renewable energy sector has seen a number of changes in the last couple of years, including changes to the broader policy and financial environment. There are many competing pressures on the industry that are having an interesting impact. Recently, Bloomberg New Energy Finance reported that the cost of solar PV had fallen by nearly 75 per cent over the last four years. A home-owner can now have a 2 kilowatt PV system on their roof for less than $4000. The NSW and QLD feed in tariff and the national solar rebate were more successful at getting solar PVs on roofs than anyone expected to the extent that both were stopped, but stopping the rebate and feed in tariff had even greater impact over the short term in getting PV on rooftops. An interesting impact of the level of solar PV now installed, is that the reduced demand is one factor keeping the daytime wholesale electricity price low. One follow on effect from that is the impact on investment decisions for large scale renewables since they must compete on the wholesale market price, and not offsetting a retail price, or net metering as homeowners can do.

Solar PV is by no means the only technology that has seen technology cost reductions. Our wind generators are setting the bar high by generating power with a market clearing wholesale price around $100 per Megawatt hour. The report on levelized cost of electricity between 2012 and 2050 released last week by the bureau of resources and energy economics shows that Renewables are rapidly becoming the cheapest source of energy Quentin Grafton, Bureau of Resources and Energy Economics said "Australia will experience an energy transformation over the coming decades that will have a profound impact for electricity networks, how energy is distributed and on Australia's ability to meet its targeted greenhouse gas emissions reductions." How true this is. The levelized cost of generation modelled and reported on has certainly dropped and is projected to continue to drop with learning curve, increased manufacturing competition and economies of scale. Putting wind and nuclear below coal fired generation in the next 20 years. But there are some big challenges required to achieve this. Large scale integration of wind puts new demand on the network due to the unsteady nature of the output. Solar PV can have much the same effect.

Solar thermal generation is technically enticing particularly with some storage capacity, but the cost of generation limits its uptake for now. The economics of solar thermal energy appear to have a sweet spot at about 2–3 hours of energy storage. This reflects the strength of the technology. Solar thermal storage could technically be extended to cover all of the night to enable plants to be dispatchable base load generators. However, the peculiarity of this is that over-night, off peak power is the lowest wholesale price. So, the market cannot support an expensive technology to supply its lowest price product. Here is yet another example of the need to think outside the silo: We have a technical solution but still a market problem.

Whilst great research, business development, tightening of technology prices continues, we've also experienced a tightening of available funds for investment globally due to continuing financial uncertainty coming out of Europe. There are renewables projects which have great potential but there is little investment capital available at present, and what exists is very risk adverse. The government is interested. With the roll out of the Clean Energy Future program, the new renewable energy board that I am a member of, ARENA, has been set up to make decisions, to fund projects, programmes of work, to assist the development and commercialization of renewable energy for Australia. There are many ways to do this. The Grattan Institute's most recent report, argued that there needs to be a greater focus on market mechanisms and less on the traditional government role of technology support, perhaps because of current market uncertainties. ARENA is approaching the strategy question with an open mind—what pathway best achieves the target of large scale, commercial uptake of renewable energy? Australia's renewable energy resources are the envy of the world: from wave energy, hot rocks, biomass and of course some of the best solar radiation anywhere, as well as an abundance of opportunities to harness the wind on land and on the sea. The challenge for this generation of engineers is how to harness these resources, working with the market, policy, national and international factors to achieve real progress and change in this area.

This brings me to another aspect of diversity. Diversity of experience to enable technical engineers to lever their expertise by building an understanding of markets, social and political impacts on these changes. I used to consider that my best efforts were spent focusing on what I did best, the hard maths and physics, the toughest technical challenges and that the political, social arguments could be left to the politically and socially minded. Experience has taught me that this is not completely true. Engineers with the training, experience and skills in understanding, resolving, and explaining the toughest technical challenges absolutely do need to engage in our social and political arguments. In my interactions on national projects and national boards, I bring of course another aspect of diversity—a Tasmanian viewpoint on issues. I identify as Australian so I don't always pick this as a diversity issue. Though whenever I fly out of Tasmania and land in Melbourne, Sydney or Brisbane at night I am reminded of how different we are. The view of the city lights in each case reminds me of just how different the perspective of the big mainland cities can be. The counterbalance to the perspective Tasmanians bring to offshore interactions is the diversity of culture and experiences personally gained through travel. This is another aspect of building personal diversity. Often Tasmanians need to travel just to completely understand how special our home is. Others travel here from interstate and overseas and realize the uniqueness as soon as they step off the plane and fill their lungs with Tasmanian air.

Tasmania's environment and natural assets are special and unique. There is no denying that, and the clean air, clean water, open spaces and community are why many of us live in this wonderful place. However, in some respects we have those assets by accident. Because we are so far from major population centres. Because it is colder than Sydney and more isolated than Melbourne. Because we have a small population and our developments have been small and haven't challenged our environment too greatly. I have seen some other states market their food, wine and scenery as clean, green and beautiful. Other places have much more sophisticated water and waste treatment, make better use of their resources, demand greater standards of energy efficiency, housing developments and industry. What I am saying is that we can't depend on our remoteness alone to secure our future. Tasmania does not have a sustainable economic future relying on current environmental assets without development. What we need to do is undertake development that complements and expands our assets and provides jobs and wealth to the state.

There are great businesses operating in Tasmania, small and moderate. Many are quietly doing their thing. Unfortunately many have found it is easier to conduct projects on the mainland. I do. My consulting business is about 50% Tasmanian and 50% mainland based. How do we build the culture that will allow us to sensitively and sensibly build the development that we need in Tasmania? Education is key. Good quality education. Tasmania is too far behind, on average. But we have talent and in my travels I find great Tasmanians all over the world. Well respected and doing a great job. I hear people talk about the brain drain. About how do we keep young people here? I disagree. Tasmania is isolated, a silo, and our tight knit community that I value so deeply as a parent, is not useful for young adults. After a great education I encourage our young people to travel. Particularly to work interstate and overseas. The challenge is: make sure they come back. And that they bring back their diverse, expanded, creative selves.

I was fortunate to have the great opportunity to study at Oxford University and work on cooling systems for Rolls Royce jet engines. Working and studying overseas I found to be a great way to build my own diversity of experience. To be immersed in a culture, to try to work in teams, to influence and collaborate with people from other cultures was for me a way that I truly started to understand some of my own culture, and to understand what it meant to think about things in a totally different way. For example, while studying at Oxford I met a German student and we started talking about Tasmania. Oh, she said, you are Tasmanian, you people were so cruel to the aborigines. That actually stumped me for a moment, since all the way through school I had learnt about the English settlers and how cruel they were. Never did I truly identify as one of them. Of course now that view has more widely become part of our Australian, psyche. It was a learning experience for me at the time.

Ok, so great- I had diversity of experience gained from the opportunity to travel and I worked at Oxford and learnt to see through different eyes so what does that mean for Tasmania? Well, when I travelled and worked I made contacts, a network, a reputation for myself as someone who had skills that were in demand. So, I suggested that Rolls Royce fund my research in Tasmania. And with an exchange rate of $3 to the pound it was an offer they couldn't refuse. But I wasn't the first Tasmanian to do that. There was Greg Walker before me, and Arch Oliver before him. Each of us dedicated Tasmanians had to leave the state, work overseas, make connections and then travel regularly since it is a rare event for the world to come to Tasmania. But the task now is so much easier than it was and getting easier, with Skype, email, phone and air travel. Clean, efficient energy is a global challenge. Tasmania has unique skills and experience in this sector, but we need to build our networks, knowledge and reputation by going out and coming back. Regularly, as I outlined earlier, renewable energy technologies are now undergoing the transition from a side issue, to becoming a large and serious contribution to the global energy market. My experience in working with Hydro Engineers and my background in gas and coal has helped me, because they have all been about big, serious, traditional engineering projects. Now I am bringing all these technical, market, strategy and governance skills together as Executive Director of Rainbow Bee Eater. Project Rainbow Bee Eater is developing a system of nodes or small power stations which will produce gas, electricity, heat and biochar from wheat straw, initially in the wheatbelt region of WA with expansion nationally. This project is significant for Australia in developing new jobs in regional Australia, expanding the electricity infrastructure and environmental benefits of renewable energy and soil improvement with biochar. There are many projects like this in Australia, across a broad range of industries.

I have shared my thoughts on: Diversity of thought through diversity of gender; Diversity of industries which can bring a variety of solutions; Diversity of experience which bring in the needs of the market, politics and society to solve global challenges. The world and its opportunities are not going to come to us. We do need to make the effort. We need to be the ones to travel. To leave Tasmania from time to time to broaden experience but we need to create the opportunities for our young people to come back. We need to encourage the use of technology to keep in touch. Once that effort is made, we do have the advantage in Tasmania. It is easier to be creative here, where there are fewer distractions and there is not as much competition. Just look at the leadership we hold in political and environmental debate. It is easier for us to work in diverse teams and cross sectors because there are fewer of us. I call on you to build diversity in your own skills and experience, build diverse teams and encourage your teams build their experiences. Most of all I call on engineers to be involved in all aspects of the challenges and changes we have ahead. Influence the world. In the words of Theodore Roosevelt: We have too much to do to sit on the sidelines.

4.20 Lessons from History

What can be learned from the history of the Faculty of Engineering at the University of Tasmania which is of relevance to present and prospective professional engineering higher education institutions in Australia and elsewhere?

First, the process which has led to the present undergraduate course and facilities has been one of parochialism (Northern vs. Southern Tasmania) and individual's ambitions and "patch guarding."

Second, internationalization has cultural, financial, and academic standards implications.

Third, the expansion of technical knowledge in the post-computer decades has resulted in increased specialization within the undergraduate courses.

Fourth, as will be further evidenced in the following chapter, the University of Tasmania fostered the rewarding careers of many exceptional engineers and leaders.

References

Australia. (1976). *Report of the committee on post-secondary education in Tasmania.* Retrieved from: https://catalogue.nla.gov.au/catalog/2375653

Chalmers, D. R. C., & Meikle, B. (2017a). *Voices of a University: Celebrating 125 years at the University of Tasmania.* Retrieved from: https://125timeline.utas.edu.au/timeline/1900/silver-lining-mining-school/

Chalmers, D. R. C., & Meikle, B. (2017b). *Voices of a University: Celebrating 125 years at the University of Tasmania.* Retrieved from: https://125timeline.utas.edu.au/timeline/1980/professor-alec-lazenby-ao/

Doe, P. E. (2020). *Re-imagining Sandy Bay.* Submission to UTAS public consultation. (Unpublished).

Eddy, W. H. C. (1961). *Orr. Jakaranda.* Publishers Pty Ltd.

UTAS. (1955). *Minutes of meeting of the UTAS Council on 22 February 1955.*

Zeehan School of Mines. (1896). *The Mercury (Hobart, Tas. : 1860–1954). May 19*, p. 4. Retrieved 17.12.23, from http://nla.gov.au/nla.news-article9324097

Chapter 5
Graduate Outcomes

Abstract This chapter is a compilation of presentations by UTAS—trained professional engineers who graduated in the 1940s, 1950s, 1960s, 1970s, and later. These presentations were first made at an online session in October 2020 held to celebrate the 100th anniversary of the first meeting of the UTAS Faculty of Engineering. Graduates were invited to reflect on time as undergraduates and the variety and eminence of their subsequent careers. Also included are biographies and "pitches" from UTAS Rhodes Scholars. Besides the broad nature of the curriculum, the relatively small size of the faculty and the expertise and dedication of the lecturing staff, particularly Professor Oliver, are common themes of the graduate's recollections.

5.1 Introduction

To celebrate the centenary of the Faculty of Engineering at the University of Tasmania in 2021, graduates were invited to prepare and present a short one-page written, or 4-minute audio, account of their time as undergraduates and their subsequent careers. Two graduates from the 1940s responded; both were Rhodes Scholars from Tasmania. There were four accounts from the 1950s graduates, 19 from the 1960s graduates, and nine from the 1970s and more recent. Some refer to the undergraduate training they received in the School of Engineering at the University of Tasmania, and all have attained eminence as professional engineers locally and internationally.

Remarkably, Alan Stretton, who commended his engineering degree in 1944 and gained a Rhodes Scholarship to complete a MA at Oxford and is now in his late 90s, is still contributing papers on management to an international journal. Alan contributed the following "pitch" to be included in this book. Since Alan's time, there have been a further six Rhodes Scholars from the UTAS School of Engineering. The contributions from Stephen Gumley and Jane Sargison are presented in Chap. 4. Contributions from Bruce Cole, Michael Wagg, and Keir Pierson appear below

P. E. Doe, *Prospects for Professional Engineering Education in the Asia-Pacific Region*, Education in the Asia-Pacific Region: Issues, Concerns and Prospects 74,
https://doi.org/10.1007/978-981-96-3197-1_5

together with Sergio Giudici's eulogy reprinted with the kind permission of his widow, Rossalyn Giudici.

Alan Stretton (1948)

Engineering served as an entrée for me to become increasingly heavily involved in a variety of broader fields of endeavor.

I don't have any particularly noteworthy memories of my time at the university (1944–1947). I did recognize that the staff were often very hard-pressed in the late war and the early postwar years but still managed pretty well. My own interest at the time was active involvement in heavy construction, and I worked with the HEC at Liawenee in 1948, before taking up my Rhodes Scholarship at Oxford for 3 years. I got my MA in mathematics but in doing so came to recognize that I was somewhat better suited to engineering. So, in late 1951, I returned to Australia and joined the Snowy Mountains Authority. I spent most of my 7 years there in the Field Construction Branch—mainly on day labor construction, opening up roads and townships. This was followed by 3 years in Dunedin, New Zealand, working mainly as a consultant on subdivisions, followed by 2 years in Victoria and NSW with Thiess contractors, mainly on construction planning and tendering. Although this was rather mundane, it gave me a variety of different construction-related experiences.

I joined the Lend Lease subsidiary Civil & Civic in late 1961 and early on introduced network analysis planning to the company's projects, initially with CPM and then PDM in 1963. Also, in 1963, Lend Lease acquired the L.A. Allen professional management program, and I managed the whole of the in-house Lend Lease management education program for 7 years from 1965. So I came to see myself as a manager, albeit in an engineering-related environment. Lend Lease also introduced an in-house marketing program from 1965, which added a customer needs focus to my previous engineering product focus. My involvement with CPM and PDM eventually led to an increasing interest in project management in broader contexts than construction. I was the guest speaker at the first public meeting of the Project Managers Forum in 1976 and was actively involved with its successor, the Australian Institute of Project Management from 1989 (elected a Life Fellow in 1996). I became similarly involved in North America's Project Management Institute and was director of its Standards Committee (responsible for the PMBOK Guide) from late 1989 to early 1992. This, in particular, introduced me to a mind-boggling range of different project types and application areas.

My 27 years with Civil & Civic and Lend Lease illustrate my increasing involvement in broader fields of endeavor. These included development and implementation of new project planning and cost control systems, management of R&D, long-range company and corporate planning, management education, organization development projects, employee benefit initiatives, corporate outreach projects, many troubleshooting missions, and a good deal of counseling. From 1968 to 1975, I also undertook part-time teaching of project planning in graduate diploma and master's courses in the School of Building at the University of New South Wales and did a variety of other external stints in part-time education, course advisory committees, and other external activities.

In 1988 I was recruited by the newly created University of Technology, Sydney (UTS), to develop and deliver a new Master of Project Management program (later augmented by graduate certificate and graduate diploma courses and a doctoral program). The course was well received, and the strength of the support staff allowed me to reduce involvement to the point of retiring from UTS in 2006.

I had started writing and presenting papers on project-related topics from 1977 and continued to regularly publish articles. To date, these numbers well over 270, many of the more recent ones being published in the monthly email journal *PM World Journal*.

In 2005 I was awarded an honorary PhD in strategy, program, and project management (ESC, Lille). This was essentially a peer recognition of contributions made over the years in these sections of the diverse areas into which I had unexpectedly progressed from an engineering degree over 70 years ago. I suspect that many engineering graduates will have had experiences not dissimilar to mine. This is one of the reasons I am offering the above, which I hope might be useful.

Bruce Cole (1953)

I started engineering at UTAS in 1949 at the age of 16 1/2 having received approval to come before the age of 17. There were only six engineering students doing the first year with me: Ken Briggs, Allen Trail, Ken Hughes, Gordon Jakes, and Peter Jefferies. There were some return servicemen ahead of us. Maths and chemistry lectures were in the old physics building on the domain. The university had started occupying the Sandy Bay rifle range site. Physics and geology were in temporary weatherboard buildings at Sandy Bay. Mechanical engineering lectures were in the Tech College in the mornings together with Engineering Drawing and Design, Workshop Practice, and other practical classes in the afternoons. The students' common room had three rooms: one for men, one for women, and a middle room which had a table tennis table and the door at the back which the caretaker's wife converted into a servery selling drinks and snacks. That was the university cafeteria. The main campus also had a netball court and tennis court both currently car parks.

Arch Oliver lectured us in materials, structures, and hydraulics. He had created a small structures lab with hands-on models and a huge testing machine. There was no hydraulics laboratory. There was an electrical laboratory in the physics building basement. There was a steam engine in the Tech College basement for mechanical engineering. The first 3 years of engineering were the same for everyone. Early in the final year, we chose appropriate subjects for civil, mechanical, or electrical engineering. I chose civil. Professor Burn was head of the department and gave us a few lectures in structures. He and Arch Oliver were great lecturers.

The course and the faculty were first class. In my final year, Arch Oliver compressed all the final lectures and lab work into two terms so that I could depart for Oxford on a Rhodes Scholarship in September 1983. The university sent exam papers to Rhodes House. In my time, there were virtually no student cars at the university. Returning from overseas in 1958, I joined the Hydro-Electric Commission which was very busy building new power schemes to meet the rapidly growing demand for electricity. Led by Alan Knight, the Hydro was assembling a very competent professional team. New sites were being investigated. Roads, dams, tunnels, pipelines, canals, power stations, and transmission lines were all on the drawing board. Construction was by day labor, the Hydro's own workforce. I started in the dam design section. In 1960 my job on-site at Catagunya Dam was to supervise the installation of 412 prestressed steel cables which anchored the concrete dam to its foundations. Catagunya Dam was the second dam in the world to be designed for anchorage by prestressing. In 1965 I was the lead designer on the 49-meter-high double-curvature arch dam at Repulse, the Hydro's first arch dam. In 1974 I spent a year in Bangkok as a dam expert advising the Royal Irrigation Department which had several earth dam failures due to the unfortunate use of dispersive soil. In 1976 I became responsible for the design of all the dams in the Pieman scheme. These were concrete-faced rock-fill dams in which the Hydro became a world leader. The highest was the 122-meter-high Reece Dam. Other positions with wider responsibilities followed. In 1989 I became project director of the King and Anthony schemes, the last two hydro developments. I was responsible for completing the schemes within budget and on schedule, a challenging task as we had to run down the workforce at the same time. Retiring in 1993, I look back on an exciting and satisfying engineering career. The many power schemes will serve Tasmania well for many decades to come.

Sergio Giudici (1960)

Dr. Sergio Giudici arrived in Tasmania with his mother from the Italian Alps at the age of 10 to join his father who was working with the Hydro-Electric Commission at Butlers Gorge. The young Sergio won a prize for English after just 1 year at Butlers Gorge Primary School.

Sergio completed his secondary education at the Hobart Technical High School, now New Town High School, matriculating in 1955, and went on to complete an honors degree in civil engineering at the University of Tasmania in 1960. In the same year, he was the first migrant (new Australian) to be chosen as Tasmanian Rhodes Scholar to study at the University of Oxford, gaining a doctorate for a thesis on the buckling strength of framed structures.

From that notable beginning in his engineering education, Sergio joined the then Hydro-Electric Commission in 1963 at the age of 25 as an engineer in the Dams Department.

He stayed with the Dams Department until 1977, by which stage he was senior engineer. He helped design the Cethana and Mackenzie concrete-faced rock-fill dams and was principal designer of the Gordon Dam. The 140-meter double-curvature arch Gordon Dam is one of Tasmania's icons and the structure for which Sergio gained most fame.

In 1977, he was promoted to head the Structures Department. He directed and helped to design the Pieman Road, Murchison and Mackintosh Highway bridges, and the Mackintosh, Bastyan, and Reece power stations. These power stations incorporated an innovative design that shortened construction time and became standard for all subsequent Hydro Tasmania power stations. Since 1964 with the commissioning of the joint UTAS-HEC Computer housed in the physics building on the Sandy Bay campus, computers have been increasingly used in engineering design. Sergio was at the forefront of developing and using microcomputers and software for complex stress analysis and other tasks including the analysis of seismic data for predicting earthquake loadings on the Tasman Bridge.

In 1983, Sergio was appointed Chief Engineer of Design Group One where he managed a group of 18 engineers and scientists involved in hydraulic, geomechanical, geotechnical, mechanical, and technical work on the King, Anthony, and Gordon power schemes and the Crotty Dam.

Continuing his professional advancement, in 1988, Sergio was appointed manager of the Civil Investigation and Design Group. Here his responsibility was for the design output of 50 engineers grouped into specialist departments for design, feasibility studies, and maintenance of dams, tunnels, structures, gates, hydraulics, roads, and future hydroelectric schemes.

In 1991, Sergio became the founding general manager, Consulting, where he organized and managed a multidisciplinary group of 250 people who designed the civil, electrical, and electromechanical components of Hydro Tasmania's last two power schemes and maintained Tasmania's 54 large dams and catchments. He was the driving force behind Consulting's plans and activities to market its expertise to external clients.

The name Sergio Giudici is known throughout the world for innovative design techniques in major hydroelectric structures. He was appointed by the World Bank and Asian Development Bank and an international panels of eminent engineers to review the design and safety of dams in China and Indonesia.

In the following years, he undertook assignments in Indonesia, Fiji, Kazakhstan, Lao PDR, Malaysia, and the Philippines as well as the South Australian, Victorian, and Queensland state governments.

He also lectured part-time at the University of Tasmania on hydro-engineering. Sergio maintained an abiding interest in university education, contributing in various ways to committees and departments. He had an active role with the Church earning him the respect and admiration of many clerical friends. St Mary's Cathedral was filled to capacity with family, friends, colleagues, business associates, priests, and two archbishops for his funeral. The people of his village in northern Italy were very proud of him, and, as a young man studying at Oxford, on his first visit back to his birthplace in the Italian Alpine mountain village of Sernio, his primary school teacher claimed this Rhodes Scholar and engineer as her own!

Most of all, he was devoted to his family, his wife, Rossalyn, and their seven children. He had the pleasure of working with his daughter Christina but did not get to see his grandson, Oliver, graduate with an honors degree in civil engineering from the University of Tasmania and work with the Hydro.

He was recognized in 1994 for his endeavor by being admitted to a fellowship in the Australian Academy of Technological Sciences and Engineering, a rare national honor. He is one of five Tasmanians who are fellows of the academy.

Sergio and his top people, together with the then marketing group Hydro-Electric Commission Enterprises Corporation (HECEC), "sold" engineering expertise to the world. In his valedictory address, Sergio said he was "most pleased at having been able to hang on to the skills and expertise embodied in the consulting business unit."

After leaving the Hydro and undergoing heart surgery in 2000, Sergio continued working as an independent consultant undertaking assignments in dam, hydropower engineering, and wind energy studies as an expert reviewer in PR China and New Zealand. He was working in New Zealand up to the time his heart failed in April 2002.

Dr. Giudici is considered one of Tasmania's finest engineers, and as a mark of the respect in which he is held, Hydro Tasmania decided to fund a scholarship in his name—the Dr. Sergio Giudici Hydro Tasmania Research Scholarship—to encourage talented graduates to undertake research in engineering or the environmental sciences at the University of Tasmania.

As well, the Hobart City High School old scholars contributed funds to establish the Old Scholars Sergio Giudici Award commemorating the school's first Rhodes Scholar. This award recognizes initiative and all-round ability; academic achievement and sporting performance are important factors as are contributions to the life of the school, community involvement, and personal interests and activities.

Keir Pearson (1964). Distinguished Professor Emeritus Dr. Keir Pearson, BE (Hons), PhD, FRSC. (1964)

I graduated with First Class Honors in Bachelor of Engineering (Electrical) from University of Tasmania in 1964 and was awarded a Rhodes Scholarship to complete a PhD at Merton College, Oxford University.

A year after going to Oxford, I switched from engineering to physiology (more specifically neurophysiology) and got my doctorate in physiology in 1968. This was not as big a switch as it sounds because neurophysiology is largely focused on electrical signaling in nerve cells so my background in electrical engineering was very valuable to have. Most of my research over a period of 45 years was aimed at trying to understand the neuronal mechanisms regulating walking. I would consider this research as basic, since it had no immediate practical applications. Nevertheless, it was of interest to engineers building walking machines and to clinical researchers trying to figure out procedures for enhancing walking in patients with spinal cord injury.

After 5 years at Oxford, I moved to Canada to take a position at the University of Alberta. The freedom for research at the university in the early days was enormously beneficial and gave me many opportunities to follow my interests. With very talented colleagues, I had the opportunity to establish an internationally recognized research center in motor control neurophysiology. The close proximity to the USA was a huge asset, and this was definitely a factor in my not considering returning to Australia. I also got to appreciate that Canada is a wonderful place to live (despite the winters). I have been happily retired now for 12 years, and it seems I will spend the rest of my time living in Edmonton. I do get back to Tasmania now and again. My wife and I visited my brother in Port Sorell last November, and we also had a chance to visit family and friends in Hobart—definitely evoked a lot of nostalgia for Tasmania and memories of the great start the University of Tasmania gave to my career.

Michael Wagg (1978)

Michael's heritage is from the Huon Valley with both his parents growing up on apple orchards before moving to Hobart. Michael attended both New Town High and Elizabeth Matriculation College. He started his engineering degree at the University of Tasmania in January 1973 commencing like many others of that time with the tech drawing course undertaken before the regular term commenced. He completed a combined Bachelor of Science and Bachelor of Engineering with Honors at the university. He took up his Rhodes Scholarship in 1978 completing a Doctor of Philosophy in Engineering at Oxford University. Michael's academic focus was in electrical and communications engineering. At Oxford, he investigated the propagation properties in crystals, of sound waves generated by electrical transducers. The aim is to optimize the performance of the filtering capabilities of the crystals for use in applications such as radar and television systems.

Michael had good intentions of pursuing a research career when he left Oxford in late 1981. A job offer at the CSIRO at Linfield was an exciting prospect. However, hard economic times meant that the then government axed many positions including Michael's exciting career prospect at the CSIRO during his 2-week return journey from the UK to Australia. This government induced fork in his road which meant that Michael pursued an entirely different career opportunity as he accepted a role at OTC in January 1982 in the then National Satellite System Division to work on the development of Australia's first satellite communication system. The medium was different with a satellite nearly 36,000 km above the Earth communicating through the vacuum of space and the Earth's atmosphere. However, the sound engineering fundamentals obtained at the University of Tasmania and the modeling and research skills developed in Oxford were used to support the development of the models used in the design and building of the satellite system. In 1983, Michael was one of the first employees at Aussat, the government-owned company, set up to own and operate the Australian National Satellite System.

Michael had a variety of technical and management positions at Aussat including being the manager of the team that developed a pioneering world's first satellite land-based mobile telephone system. The team started with a blank sheet of paper undertaking everything necessary to develop, test, and perfect the technical specifications of the system as well as set up the commercial requirements for the subsequent service. This included contracting with third parties for the building of all of the components of the system from the mobile satellite telephone user terminals through to the satellite systems in orbit at 36000 km as well as interconnection of the system to the global telecommunications network. During this period, Aussat was sold to the new telecommunications company Optus in the early 1990s. Michael stayed with Optus for another 25 years in a variety of technical and managerial roles which included the development and contracting of a joint communications and defense satellite and the purchase of multiple satellite launch vehicles, aka rockets.

Michael's career has been in the satellite and telecommunications industry; however, he has also dabbled in university life as an industry advisor and guest lecturer at a number of universities across Sydney.

Michael met his partner Jo (a fellow Tasmanian: quality attracts) in Sydney, and they have two children, two granddaughters, and a grandson. Their daughter's 3 years in London from 2014 was the excuse for frequent trips to the UK including enjoyable stays with old friends from his days at Oxford. He and Jo still have family in Tasmania and regularly visit the state.

5.2 1950s Graduates

Gregory Leonard Crew (1959)
I started the engineering degree course in 1955 after gaining my Matric from Hobart High School. I graduated in 1959 with the BE (Hons) degree.

I was fortunate to get a PMG cadetship, which paid a modest allowance and provided work experience in the vacation periods. This led to a career in telecommunications, first in the PMG Dept., later Telecom Australia. I moved from Tasmania to Melbourne with PMG HQ in 1963 and was involved in the introduction of new telephone exchange systems providing direct long-distance customer dialing interstate and then internationally. In 1975, with the formation of Telecom Australia, I was appointed superintending engineer of switching design and then assistant chief engineer in charge of the Australian telecommunications construction program. I gained a M. Admin degree from Monash in 1979 and was appointed general manager, Commercial Services, in 1981.

In 1983, I accepted the position of general manager, Engineering, with Hong Kong Telephone Co., subsequently becoming managing director of both Hong Kong Tel. and Cable & Wireless HK Ltd. I served on the Council of the Open University of Hong Kong and was honored with a DSc. (c.h.) in 1999.

In 1993 I moved to London with C&W Hong Kong as deputy chief executive and COO of Mercury Communications. I retired home to Melbourne in 1995. Over the next 10 years, I served as the first board chair of ACIF (the first telecom self-regulating authority as competition was introduced) and on the boards of a number of companies, including as a founding director of ICANN (international Internet domain name authority) and AuDA (Australian .au domain name authority).

At UTAS, I remember:

- Professor Archie Oliver, abusing us as we attempted to mix concrete for slump tests: “Haven’t you seen your mother mix cakes?”

- Professor Gordon Newstead in Elec. Eng. (I subsequently replaced his brother, Tony, as head of Engineering, Hong Kong Telephone Co.).
- The infamous Prof. S. S. Orr scandal, which reflected glory on no one. Very interesting to us innocent undergrads.
- The first parking meters in Hobart. We paid our money and held a picnic on one—"6d and they're half (h)ours." The crowd disrupted traffic. The police were not amused.
- The first roll-on ferry between Tassie and Melbourne. Having borrowed a corset, I was the "roll-on fairy" in the annual students' parade. Do they still hold them?
- Rugby with coach "Tiger" Wilson.
- Travel between Domain (Engineering) to Sandy Bay (Science Faculty in Musset huts) and Hobart Tech. (drafting and workshop practice). My mate, Bob Purdon (sadly deceased), had a car—he was very popular.
- Student Union dances—I met my wife there! We've been married for 60 years!
- Surveying around the Domain, using five-figure log tables (no computers in those days). Studying electronics using vacuum tubes. I used early transistors in my Honors project.
- I was nearly electrocuted in the Elec. Eng. Lab. when someone forgot to place a switch in the 400 V DC power circuit. I survived, surviving still.

Roderick Cuthbert (1954)

My most challenging project as an engineer.

In 1960 the Hobart City Council ran a competition for a design for a water fountain to be built in the newly completed Railway Roundabout at the junction of Liverpool St and Brooker Av.

With two other colleagues at Cadbury Confectionery Claremont, a design was submitted and, lo and behold, it was judged the winner.

I had thought our involvement ended there. Not so. The city architect Bert Dechaineux then contacted me and informed me he now looked to me to manage the construction of the fountain.

The media had suggested our entry was a Cadbury exercise, and he suggested the public saw it as such.

I had good connection with many Hobart contractors and figured I could look to selected ones to do the job. However, try as I did, I could not find a consultant who would tackle the hydraulics involved. So, the Cadbury directors having told me to proceed, I had to tackle the rather complex piping design. There were three rings of jets in the top bowl, and all these had to be individually fed from the water circulating pump and controlled to varying heights.

My first problem was establishing the water flow from each ring which varied with jet heights. In a moment of sheer luck, I decided I may be able to call on the eighteenth-century mathematician and scientist Isaac Newton who had developed a number of equations relating to objects in flight and subject to accelerating force: V squared = U squared +2as. V is the velocity at the end of a travel s with starting velocity U and acceleration rate of a. I rearranged this to apply to a fountain jet and found on running a test with a hose, measuring tank, and stopwatch Isaac had got it right!

I then proceeded with the very complex design which also included control systems. The fountain was launched by the Governor of Tasmania, Sir Stanley Burbury, on what was, fortunately, a very calm evening. Within a week of start-up, the wind control, which one must have on all tall jet fountains, proved useless in the low wind speeds involved.

I now had the council on my back to fix it and the media commenting. Without going into detail, the anemometer I developed relied on very light spherical ball suspended on a wire conductor which passed through a hole in a brass plate just above the ball. When the wind moved the ball, contact was made with the brass plate, and this, through a 12 volt circuit, signaled the controls to reduce jet height.

I was retained to keep an eye on the fountain but no further problems.

In my role at Cadbury, I encountered many taxing design problems but none with my efforts watched by public and media.

Kenneth John Hughes (1953)

I began engineering at UTAS in February 1949. There were only six students doing first year with me, Ken Briggs, Bruce Cole, Gordon Jacques, Peter Jeffries, and Alan Traill. There were several students ahead of us, including several returned servicemen from World War II. We started in February to commence 3 weeks of full-time Engineering Drawing and Design 1.

I began working for the Hydro-Electric Commission of Tasmania on January 4, 1953, at Bronte Park which was a construction village entirely owned by the Commission in the highlands of Tasmania about 160 km NW of Hobart, adjacent to the left bank of the Nive River.

I worked on the supervision of the following structures and facilities: Pine Tier Dam, Bronte flume and canal, and the Clarence steel and wood stave pipeline (an additional supply of water) and also the concrete batching and mixing plant for the whole of the Bronte and Lake Echo area (the main aim in the concrete mixing plant was to achieve adequate strength, durability, and workability).

Two incidents which occurred during the construction of the Bronte Canal were two landslips of the shale bedrock into the partially dug canal. These had to be remedied by modifying the alignment and constructing a reinforced concrete flume to replace the canal, in each case for about 300 m.

Later on, I worked on the Echo earth dam which provided the main storage for the Tungatinah and Echo power stations. One difficulty with the construction of Echo earth dam was the supply of a satisfactory material. Initially, a relatively plastic material was used, but it proved difficult to dry to optimum moisture content in the prevailing climate which included two unusually wet summers (compaction can only be achieved in summer). Eventually, a more appropriate material with a courser grain was found at a higher elevation within about 2 km which was easier to dry and compact. When I revisited these sites in 2016, i.e., 60 years after construction, they all appeared to be in good condition.

In January 1956, I transferred to the HEC head office in Hobart to work on the Derwent Pumping Station and pipeline which was required to pump 3 m^3 per second into the Tarraleah No. 2 Canal. The main issues were the stability of the reinforced concrete pumping station during normal operation and especially during the occurrence of the maximum probable flood in the Derwent River, the reinforcement required for the boxlike structure, the trash-rack structure, the concrete anchor block for the pipeline, a nonreturn valve, and an enclosed surge chamber.

In the years that followed, I had the opportunity to work for the Electrowatt Company Zurich, Switzerland. I did the surge tank calculations for the Misoxer Power Development and the steel-lined pressurized shaft. I also worked on a number prefeasibility studies, including of the Homathko River in Canada, the Congo River in Africa, and a pumped storage project in Europe and worked with Sandwell & Co, in British Colombia, in the pulp and paper industry on structural work and mill alterations.

In 1963 I returned to Australia and worked with Snowy Mountains Hydro-Electric Authority on upper Snowy and upper Geehi and Tumut 3 projects.

In 1973–75 I was working on the feasibility study of the Cirebon-Cimanuk irrigation and hydropower project in Java and in 1975–1976 for the design of this project. In 1977–1979, I was project manager in Nepal for the Gandaki River Basin Study which includes about one quarter of the area of Nepal, funded by UNDP (United Nations Development Program). This work was followed by three feasibility studies: Gandaki "A" hydropower development, fieldwork for Marsyangdi hydro, and Bagmati hydropower and irrigation protect in the south of Nepal.

Toward the end of my career, I worked on the feasibility study for the Apsley River pumped storage scheme for the Electricity Commission of NSW and on an assessment of the hydropower potential of the whole of Papua New Guinea. I studied for a Graduate Diploma in Engineering Computing and Applications at the Ballarat College of Advanced Education. Finally, I worked relieving the quality assurance engineer on the supervision of construction of the Jugiong Bypass to greatly improve the alignment, grade, and capacity of the Hume Highway with a dual carriageway.

5.3 1960s Graduates

Chris Close (1966)

Recollections from Time at UTAS

I have of course many memories of my time at the University of Tasmania, 1962 to 1966. The first is of receiving a letter at home in Queenstown in late January 1962 advising that engineering students were required to attend an introductory engineering drawing class in February. This was news to me as I was not planning on commencing my engineering degree until March. Nevertheless, I took the bus to Hobart a few days later and fortunately was able to move into Hytten Hall almost immediately.

Hytten Hall was my university home for the next 5 years, and it was a great place to stay. George Wilson was college warden combining some of his experiences from Cambridge University, including the requirement to wear undergraduate gowns to dinner, with a reasonable strict study regime and a very liberal approach to social events. I well recollect the college barbecues and the excessive, to say the least, beer consumption, the parties, and the water fights! George also found time to lecture in history and coach rugby.

The 1960s was prior to the introduction of the HECS scheme, and many of us were assisted by scholarships offered by a range of government and private organizations. Some of these scholarships involved future employment obligations, and some were just merit based, while many of our fellow students were from overseas and had Colombo Plan scholarships. In my case, I got by very well with a scholarship from the Mt. Lyell Mining Company, which was offered to the children of employees, and a Commonwealth Scholarship which covered university fees. Additional income was also earned in the summer holidays when it was requirement to have engineering-related work experience.

The engineering drawing class mentioned earlier was intended to provide an introduction to engineering design, part of the civil engineering course. I don't remember much about the subject but am reminded of it by the magnificent Wild drawing instrument set that I used at various times during my career and still use occasionally.

Career Summary

- *1966 to 1968: Hydro-Electric Commission Tasmania.* Design of construction equipment and methods for tunnels and dams on the Mersey-Forth Scheme.
- *1968 to 1976: various contractors in Tasmania, Queensland, and the UK.* Site engineer on projects including the Mersey River Bridge, silos in Devonport, a nuclear submarine base in Plymouth, UK, and piling contracts throughout Queensland.
- *1976–1989: Fluor Australia.* Cost estimating and control for a number of coal projects in Queensland, the Olympic Dam Project in South Australia, and numerous feasibility studies.
- *1989–1996: Fletcher Construction Australia, estimating manager.* Major projects won included the BHP Rolling Mill Upgrade at Whyalla, Victorian Hospitals Co-generation Project, Black Rock Sewerage Treatment Plant at Geelong, Eastern Freeway Bridges in Melbourne, and Spreads 1 and 2 Moomba to Sydney Pipeline.
- *1997–2008: Close Estimates.* Provision of estimating services to contractors for road, rail, pipeline, and materials handling projects.

Kevin Cox (1973) (transcript of video presentation October 31, 2022)
I'm a couple of couple of weeks off turning 80, and I'm still practicing engineering full time but not in a way that the engineering faculty of my time would recognize because there were no computers in Tasmania in the late 1950s. However, since graduating in the early 1960s, I've thought of myself always as an engineer as I use, in practice, the thinking tools that I learned at university. We were lucky generations; we were too young for the World War II and too old for the Vietnam War and conscription. Those of us privileged to go to university knew we would find a job when we when we graduated because there were plenty of jobs available at the time. We were lucky to go to the University of Tasmania and live in Tasmania.

The university had some great teachers; those that I remember of course were Archie Oliver but also Professor Carey in geology and "Mac" Urquhart in maths. You tend to remember the mavericks, the ones who are a little bit different, the ones who had a different view of life and were a bit eccentric I think. The people who influenced me the most were my fellow students, and I see a few on the list of today's/tonight's attendees. A couple of the highlights during my time at the University of Tasmania were the engineers and scientists taking over the student representative council. You remember that, Peter. We also held I think it was the Australian Engineering Students Conference in 1960 about Antarctica problem of the Deep South.

I've spent most of my working life developing designing and building what people now call complex adaptive systems. I founded a company that uses the idea to prove people's identity, and the systems we created now do about 50% of Australian online verifications. I left that company 10 years ago, but the last 10 years I've been working with many others to design a new economy. The present economy is one whose main objective is to generate more money. The easiest way to generate more money is to sell something that costs little to produce, and one way to do that is to monetize natural assets and trash the planet as an unintended consequence.

The new economy can actually reverse this overexploitation by producing the same goods and services with a minimum amount of money, and that's an engineering problem. One way to do that is to bring some engineering discipline into economics. Economics, when you get into it, is a very peculiar place. They build theories and pretend it's the real world, and they try and make the real world match their theories which is a very bizarre way to actually do anything. However, if we make the same amount of goods and services with less money, it actually turns out to save money, and it is economically efficient. One way to do that is to adjust this economic system to stop money tokens creating more money tokens over time, and we can solve that problem with well-designed complex adaptive systems.

I still call myself an engineer because I want to do more with less, and that's what I learned about 60 years ago.

Adrian Mercer Crawford (1968). Adrian Mercer Crawford, BE (Hons I) Tasmania 1968, PhD (Toronto), Professor, Department of Civil Engineering, University of Toronto

At the University of Tasmania, I had many excellent teachers, all of whom brought great passion to their work and who conducted the terrifying afternoon labs. In reality, our degree was like an Engineering Science degree, which is the most prestigious engineering program in many N. American engineering schools. Ours was better.

After university, I joined the Investigations group at the Hydro-Electric Commission of Tasmania. At that time, HEC had world-class engineers, many of whom I had to consult for my work. My first job was the initial design for the Murchison Dam. My next project was to produce the first designs and stress analysis of the proposed 200-m-high Franklin arch dam which was abandoned because of a World Heritage Designation. So I was off to Construction Section where I did the paper build on the Gordon Dam—that is the order in which each of the 100 blocks must be placed. The problem was that the tower cranes and the dam had a symbiotic relationship being joined by struts attached to both. Then to Gowrie Park to lead the Penstock construction group for Cethana Dam—penstock and butterfly valve installations and a penstock surface coating protection shop.

Next came short stay with SMEC, Cooma, working on a water supply for Sydney and associate Hydro works. Initially, I designed thrust foundations for large-diameter pipe bends, then the design of Wingecarribee Dam, its outlet structures, and the thrust block for the dam's radial gate arms.

In Toronto, I did my master's and doctoral degrees. My PhD dissertation was a bit schizophrenic having two disparate parts. I designed and experimented with a high-speed rock joint shear machine and developed higher order displacement discontinuity elements. Both parts had spin-offs, and the shear box was copied and miniaturized for commercial and educational use. My other work reignited interest in the DD method.

Then off to Imperial College, London, as a lecturer in rock mechanics at the Royal School of Mines. Staff include Brown and Bray. Colleagues included Burland and Bishop. Bray and Burland, like Gregory and Oliver, were exceptional mentors.

At UofT, besides teaching and committee work, I worked on the size and shape of canisters containing ceramicized nuclear waste which included plutonium. The canisters were to be placed deep in a pluton, surrounded by clay. Canister spacing and size determined whether the surrounding groundwater would boil—not a good thing.

Aside from that, I was troubled that our students at UofT were being taught to be "practice-ready." This is not the way I was educated at UTAS. Rewording Professor and President Harry Arthurs: "Being 'practice-ready' on graduation virtually ensures that an engineer will soon become 'practice obsolete.'"[1]

My career as an engineer was very fulfilling and enjoyable, due in most part to the comprehensive program of the Faculty of Engineering at the University of Tasmania.

[1] Professor Harry Arthurs, "Connecting the dots" The Osgoode Brief, Fall 2019, p. 1.

Peter Edward Doe (1963). BSc BE(Hons), PhD, FIEAust

I graduated with degrees in science and engineering from the University of Tasmania in 1963. My career as an engineer, teacher, and researcher commenced in 1963 at the dawn of the computer age. I was fortunate in getting a job as soon as I graduated with the Hydraulics Section of the Hydro-Electric Commission based in Hobart. In early 1964, the Hydro and UTAS jointly purchased the Elliott 503 general-purpose digital computer. I wrote a computer program to calculate backwater profiles and published the program as (my first) technical paper.

After a 2-year period with the Hydro-Electric Commission, I accepted a position as lecturer in mechanical engineering at the University of Tasmania, and while lecturing in a variety of subjects in all years of the BEng degree began a research program in heat and mass transfer for the award of my PhD. An interest in fish began with a 6-month period of study leave at the Torry Research Station in Aberdeen, Scotland, in 1967. There I met Dr. June Olley, who shortly afterward moved to Tasmania as head of the CSIRO's Fishery Research Unit. This began collaboration on abalone drying and led to a visit to Bangladesh in 1983 where I developed and tested the "solar tent drier." Setting up and testing the "polythene fish drier" in Bangladesh led to significant research and development activity of solar driers internationally by the Food and Agriculture Organization of the United Nations (FAO), the Tropical Products Institute, London, and DANIDA, Denmark's International Development Agency.

From 1985 to 1987, I managed a major Australian Centre for International Agricultural Research (ACIAR) program on fish spoilage in Indonesia that resulted in a significant body of understanding of fish spoilage and a generation of skilled researchers. I was able to contribute to the newly found research area of predictive microbiology by formulating a method of calculating water activity in dried fish from salt and moisture content. The name *Isohalic Sorption Isotherm* I proposed has been adopted by the international scientific community.

My work on fish drying continued with membership and subsequent chairmanship of a joint IUFRO/IUNS working party on the nutritional and functional properties of smoked, cured, and dried fish and culminated with the publication of a book in 1998 which I edited entitled *Fish Smoking and Drying: Production and Quality*.

From 1980 I led a team of scientists and engineers in a 20-year project which resulted in world-leading analysis, modeling, and practical methods for improving the quality of sawn hardwood timber—research which is still being used in the latest National Institute for Forest Products Innovation projects.

In 1996 I moved to Launceston to take up the position of head of the Department of Engineering at the UTAS Newnham campus and for a year was also head of architecture. In 2000 I returned to my substantive position as associate professor of mechanical engineering in the UTAS School of Engineering and retired in 2003. Three years later, I returned to full-time work as head of engineering at the Australian College of Kuwait (ACK). As well as leading and teaching at the ACK, I was responsible for the design and establishment of an articulation pathway from the existing diplomas in civil, mechanical, and electrical engineering to corresponding Bachelor of Engineering Technology degree courses. From 2012 to 2017, I managed the school's international engineering articulation (2 + 2) programs. In 2016 I was appointed to the full-time position as project manager of the Australian Government's Office of Learning and Teaching's strategic grant titled "Engineering Pathways for Regional Australia: viable learning platforms built by knowledge partnering."

In 2019 I joined the staff of the UTAS University College where I developed and delivered units to Associate Degree in Applied Technologies students. Recently, I was appointed to the Academic Board of the Engineering Institute of Technology (EIT), based in Perth, WA.

Tom Fricke (1968). AM, BE(Hons), FIEAust, FASCE, FAICD

Following the completion of my "very stimulating" BE(Hons) degree at the well-designed and purpose-built Sandy Bay campus of UTAS, in early 1968, I immediately joined engineering consultancy GHD in Hobart as a graduate engineer. In 1969 I was transferred to Melbourne with GHD, where I have been residentially based for most of the rest of my career.

After completing the postgraduate course in hydrology at the University of NSW, together with my wife Janet, I was then posted to Malaysia by GHD, during 1972–1974. During this time, I was working on the Pahang River Basin Project and then returned to Melbourne to be followed by numerous short stints thereafter back in Malaysia over the next 20 years.

I settled initially on a career in water engineering at GHD, heading up the Hydrology/Hydraulics Group, in Victoria and then Australia-wide. Also, two daughters arrived, one of whom is now also an engineer!

I was then appointed manager of the Water Industry Group in GHD Victoria, designing and overseeing dams, water supply systems, water treatment plants, wastewater treatment plants, pipelines, pump stations, sewers, flood studies, drought studies, water quality studies, etc., with projects located throughout Victoria and also in Tasmania, NSW, Qld, SA, NT, ACT, and WA.

In 1999, I was appointed to the Board of GHD and also to the role of Victorian State Manager, building Victorian staff numbers up from 150 to 450 over the next 5 years and focusing on all aspects of infrastructure engineering.

My GHD roles at this time also included Director of Information Systems and Director of People for GHD, during which time the GHD Business School was established. I retired from full-time employment with GHD in 2013 having served in ten different positions over 45 years of service.

Outside GHD, I have contributed as follows: lecturer in Water Engineering Design at Monash University, board member of Engineering Education Australia, Australian Government Visitor to the Cooperative Research Centre for Water Quality and Treatment, chairman of the University of Melbourne Engineering School Industry Advisory Group, member of Monash University's Monash Infrastructure Industry Advisory Council, and executive board member of the Committee for Melbourne.

Following my extensive contribution to the Committee of Melbourne's agenda over a period of 15 years, I was appointed a life member of the committee and continue to assist presently in the committee's infrastructure policy work—especially relating to the energy sector—in my "retirement."

Also, I have served as chairman of the Lonsdale Golf Club Redevelopment Project Task Force over the period 2013–2017 and continue to enjoy sailing on Port Philip Bay, playing golf, travelling, and entertaining our grandchildren!

Looking back on my school and university education, I realize how fortunate I was to be taught by wonderful mathematics, physics, and chemistry teachers at the Hutchins School, Hobart, and they inspired me to pursue the STEM path, as we know it today, at the University of Tasmania. The UTAS academic staff teaching us engineering, science, and mathematics subjects were equally inspiring and were also highly regarded both within Australia and, in many cases, internationally.

A distinguishing feature of the UTAS engineering degree in the late 1960s was the breadth of subjects taught right through the 4-year course. By way of example, my final year included electrical, mechanical, structural, hydraulics, and civil engineering subjects preceded by a solid grounding in physics, chemistry, and pure and applied mathematics. Thus, I felt that I would be able to tackle almost any engineering problem presented to me in my subsequent career, and this, largely, turned out to be the case, a great advantage in a multidisciplinary consultancy such as GHD.

A further benefit of studying at UTAS in the 1960s was the availability of first-rate engineering and scientific laboratory facilities as well as the advantage of mixing with students from other faculties such as science and law, who shared our engineering lecture theaters. The opportunities to tackle a specific research task in my honors thesis was also of benefit to my eventual career. Similarly, the compulsory 36 weeks of engineering vacation experience during the course helped to shape us as all "career-/job-ready" when we graduated.

Richard Green (1968). Richard Green BE(Hons), 1966, M Eng Science, FIEAust

Time at UTAS

One of the advantages of the University of Tasmania is the size. You work in small groups, usually two. At other universities, they work in groups of about ten both in the labs and the design studios.

When we went to uni, the Colombo Plan was in place. This meant you mixed with students from other countries. They brought different attitudes to their studies. We have been able to remain in contact with them, especially when we were working overseas. The other advantage was that we did work experience every long holiday. Most universities require only 14 weeks total. Another advantage was that we did not have to select our specialty until the final year. We therefore had experience in fluid dynamics, mechanical engineering, and electrical engineering.

Experience

I worked in all aspects of structural engineering with different materials and different building types—retail, residential, universities, schools, and commercial. This included slender 250-meter-high towers. Some of this work was done in Sri Lanka, Singapore, Indonesia, Vietnam, India, and London. This included design for cyclone and earthquakes as well as usual loads. I was a director at Taylor Thomson Whitting from 1974 to 1990 and managing director from 1991 to 2016.

Major public buildings included: Gallery of Modern Art in Brisbane; War Memorial in Canberra; Science Museum in Canberra; Art Gallery of NSW, including the new building; State Library, Sydney; five stations on Circle Line 3 in Singapore; major buildings at the University of Sydney and the University of NSW; and a number of high-rise commercial and residential towers.

Projects in Tasmania included the Derwent Sailing Squadron breakwater/Marina, Tasmania Museum and Art Gallery, and Parliament Square. I have been a board member and president of the Association of Structural Engineers and a board member of Consult Australia.

Ian Howard (1969)

I commenced engineering at UTAS in the early 1960s. Initially, I found it a struggle having not sat Maths 3 at Schools Board in Hobart nor Maths B in Matriculation. This was quickly revealed in pure maths 1 and 2, whereas applied maths with its practical concepts was not an issue. Later in second year when the engineering subject commenced, I found the theory and practical laboratory work complemented each other giving enjoyment and success.

I undertook the fourth year civil subjects with the exception of studying Hydraulics IIC with Greg Walker studying compressible fluids in lieu of geology for engineers. Later as a retiree bushwalking and kayaking after returning to Tasmania, I understand how little I know about this state's geology.

Attending my first and only engineering students' intervarsity in Perth, WA, in the mid-1960s (and had a great social time with occasional talks and field trips), I realized that UTAS was different to the larger Australian universities. Their students were narrower in their field specializing early in their course and had little contact with nonengineering fellow students on campus and somewhat less social confidence. This general engineering later served me well throughout my career.

During the course, my birthday was drawn out in the Vietnam War National Service ballot along with a disproportionate number of my colleagues. I say disproportionate as a straw poll soon showed the ballot when combined with the compulsory "medical" allowed the army to preselect skills they required. I was fortunate to be one of few chosen for officer training at a special unit at Scheyville near Windsor NSW established to undertake a world first in training young men in a 14-week course the skills to be a second lieutenant infantry platoon commander in Vietnam. This course was very demanding with long hours, strenuous field exercises, and extreme discipline. The system of weekly assessments meant 50% of my intake of 100 failed the course.

Following graduation from Scheyville, I trained in the Royal Australian Engineers in roadbuilding, Bailey bridging, mine laying, and clearing plus tunnel searching and destruction. Few of the National Service Engineer Officers including myself were not posted to Vietnam but rather were used to replace in Australia permanent officers who were part of their long-term careers. A benefit of this experience was the formal and practical leadership training looking after young men 24 h/day while they were away from their homes and families spread across Australia. This experience is not available in normal peace time employment for men in their early 20s.

I commenced "civvy street" employment with Australian Paper Manufacturers (APM) in 1971 on a large paper machine construction involving process piping design and tender letting followed by installation and commissioning at Australia's largest pulp and paper mill at Maryvale near Traralgon, Victoria. Contact law and a little accountancy from our fourth year subject Business Practice for Engineers came immediately to the fore.

After 5 years on design and construction projects in APM, a wise mentor said to me: "Ian if you want a career in this company you have to work where the money is made in production and marketing not in construction where its spent." Wise words!

Following a period in paper mill maintenance, the above words came to fruition with the mill manager position at APM's recycling mill producing packaging material for fiber board boxes at Spearwood in Perth, WA. This was like a university of papermaking and management being 4000 kms and 2–3 h from my boss in Melbourne—also a break for a young family with two children and the opportunity to meet new friends and explore that vast state.

The next mill management position was in 1981 in Botany, NSW, a really tough role with demanding customers, competition from local manufacturers and imports, and a hot bed of trade union power.

In 1985, AMCOR (APM), I made a strategic decision to branch into office papers and an exciting project with market research, large capital expenditure, overseas consulting, trials, and international equipment purchases. I was chosen to lead this project with Maryvale Mill, Traralgon, the chosen mill, because of its bleached eucalypt pulp mill.

A year working in APM's 200,000 hectare plantation estate followed which after manufacturing with its daily score card is a pursuit where most foresters who plant the trees don't harvest their results. Forestry is a highly statistical pursuit with farming on a large scale and 25-year cropping. Customer demand is hard to predict so far ahead. Forestry is interspersed with what is known in Tasmania as "the forest wars" though in Victoria and overseas it's a respected pursuit.

The following managership of all the Gippsland operation manufacturing 800,000 T of pulp and paper plus the forest estate with 1200 employees was a career highlight which also required media training and radio and TV interviews together with harnessing community support for this vital regional industry.

Attending an executive training program at Stanford University in 1995 was an amazing experience being exposed to international managers and world-renowned lecturers and mixing with young professional IT male and female staff from the surrounding Apple, Google, and Hewlett-Packard companies. The integration and cooperation of Stanford University with industry is on a scale not known in Australia.

Following this experience, a change in career occurred with a move to division responsibilities including a sales and marketing group responsible for printing and copy paper manufacture and sales locally and overseas based again in Melbourne. Again, our family of five had another move which suited our circumstance with children at secondary school and university.

Next was an experience with a Tasmanian twist—the due diligence and purchase of APPM fine paper operations in Melbourne, Burnie, Wesley Vale, and Shoalhaven Mills. These three mills progressively closed over a period of 10 years as the previous owners had spent negligible capital on equipment and cost reduction initiatives—a sad outcome but not an unusual Australian story.

My full-time career final role was an extensive supply chain project to install a business process to start at demand forecasting, order intake, manufacturing, delivering and invoicing with cost, and management accounting throughout. To facilitate this, a company-wide IT business management system was installed which is not a step I would recommend to the unprepared.

After retiring at age 58, we built a house in Blackmans Bay after never having been employed in Tasmania. We had met there at a pajama party in the UTAS common room in 1968 so it was an easy transition though now having three children and seven grandchildren in Melbourne means near monthly commutes. Retirement in Tasmania is ideal for Jean finally settling down in our 20th living space with the availability of sailing, bushwalking, and kayaking in a non-big city environment.

I continue to mentor often outside engineering using the lessons of life.

So where did the UTAS environment help?

Everyone believes their institution was probably the best, and for me the benefits were:

- Small class numbers and good laboratory facilities
- Easy access to lecturers, tutors, and helpful fellow students
- The general nature of the course providing a working knowledge of many topics
- The importance of the compulsory 35 weeks of gainful engineering vacation experience
- Mixing with other nonengineering faculty members
- The ease of movement between facilities
- The reputation UTAS had in producing well-rounded employable graduates

Phillip Mathers (1969). FIEAust (Rtr)

Memories of UTAS Engineering

In 1964, at the age of 17, I commenced my engineering studies at UTAS Sandy Bay. Fortunately, I had a Commonwealth Scholarship to help with the fees. For our first class, we gathered for Technical Drawing with Ted Sturges. Professor Arch Oliver made a brief visit and outlined the high failure rates ahead of us if we did not work hard. In this year, we also enrolled in a TAFE course covering the use of lathes and other workshop equipment. To this day, I still use the calipers, scriber, and small vise made during this course.

Lecturers that I can remember include Jack Beresford, Dr. Malcolm Gregory, Prof. Arch Oliver, Prof. Charles Miller, Bob Wherrett, and Godfrey Burrell. Memories of Prof. Oliver include him filling a blackboard with partial differential equations, his testing sense of humor, his pipe, and his wise advice about buying cars: "never invest in depreciating assets," he said and practiced.

As engineering students in the 1960s, we had the benefit of being exposed to major engineering works and related industries in the state, and this was reinforced through vocational employment during the summer months. My vocational employees included the Electrolytic Zinc Company and the Hydro-Electric Commission with work at Meadowbank Dam, the Mersey-Forth Power Development, and the Hydraulic Design Section (which included time working in the hydraulics laboratory at UTAS Engineering).

In our final year, geology was included in our curriculum. We had several enjoyable field trips led by the colorful Prof. Sam Carey and Dr. Max Banks.

Apart from study, we had a good social life while at "uni," and I was an active member of the University Basketball club and played in three national intervarsity competitions.

Career Summary and Achievements

Qualifications and Professional Memberships

Bachelor of Engineering (1969)

Postgraduate Diploma in Management, Deakin University (1991)

FIEAust, CPEng (Rtr)

Career

In the summer of 1969, I commenced work for the Hydro-Electric Commission as a construction engineer on the Gordon River Power Development. Over the next 7 years, I was privileged to be involved with the construction of this development from the very early days of construction through to its completion. I was involved with many works, including the following:

- Installation of Strathgordon Village services including commissioning of the water treatment.
- Site preparation, erection, and operation of 200 m^3/hr concrete aggregate plant. Construction of roads and site development for the switchyard and intake structure.
- Gordon Dam—concrete engineer for the initial phase of construction. The largest pours were of the order of 700 cubic meters.

- Engineer in Charge, Gordon Dam—diversion closure works and tunnel plug construction.
- Engineer in Charge, Gordon Power Station High Pressure Works—including concrete lining of the 2-m-diameter power tunnel and distributor (including cutting-edge high-pressure grouting of same) and penstock installation.

After seeking to broaden my knowledge, in 1977, I was transferred to the civil engineering design section in Hobart. Here I gained valuable experience in dams and structural design, metals technology and contract planning, documentation, and management.

In 1983, I made a major career change when I accepted the position of project engineer for the Gordon-below-Franklin scheme. When this scheme was stopped by the Federal Government, I was appointed to the position of project engineer for the King River and Anthony Power Developments (the last major hydroelectric power developments to be built in Tasmania).

In the latter part of 1992, I accepted a position in Generation Division (responsible for the management of the state's generation infrastructure). Over the next decade, I held several senior asset strategy and management roles. This work included the responsibility for major economic analyses that determined long-term strategies for several power schemes and their components, capital planning, a review of potential mini-hydro projects in Tasmania, and the implementation of advanced asset management practices.

After more than 34 years of a rewarding professional career with Hydro Tasmania, I retired in 2003. During this time, I worked with a large number of multidisciplinary teams ranging from professional employees through to site laborers. It is satisfying to know that we all contributed in our own way to the building of significant hydroelectric infrastructure that will serve our community for generations to come.

Postretirement

Following retirement, I complemented my new lifestyle by undertaking several consultancy projects in both Tasmania and Queensland.

In 2008, after encouragement from Bruce Cole, I prepared a submission for the Institution of Engineers Heritage Group that resulted in the Lake Margaret Power Scheme being awarded a Historic Engineering Marker.

In 2010, I was awarded the prestigious *John Madsen Medal* for best Engineers Australia paper in the *Journal of Electrical and Electronics Engineering Australia* for the paper "Lake Margaret Power Scheme: A Long History and Assured Future."

Mike Rodd (1968)

After completing the final exams, we had to start and think about where we were going to work, and as I recall the options that came up were Cadburys in Hobart, Boeing in Seattle, Gas and Fuel Corp in the Latrobe Valley in Victoria, and GHD Consulting Engineers in Darwin.

Well, I didn't like chocolate that much, the Briquette Factory was in decommissioning mode, and at that stage of my life Seattle seemed far away. Vacation work after third year was with GHD in Hobart on the Risdon Brook Dam construction (David Swifte as I recall), so not withstanding leaving the girlfriend in Hobart, I headed up to Darwin which was to me at the time I remember as being someplace on a map on the north coast of Australia which I found was also far away after driving up there.

Darwin in the late 1960s was a bit of an adventure, but the heat drove me to a job with the NSW Water Resources based in Tumut in the Snowy Mountains where I was working in the area of river flows, flooding, and related hydrographic activities. I thought that the climate was pretty good there, and it was where I met my wife—and we still enjoy life together.

My Honors project at UTAS was the flooding of the lower reaches of the Hobart Rivulet which I am sure had led me into the watery area of my career. It was a great lead into that area and clearly has had a strong influence. I do recall that my project commenced just downstream of the city center (Cat and Fiddle), and I thought at the time that my "inflow" was being restricted—certainly proven to be the case from recent flooding in the city center.

After a few years with Water Resources, I returned to GHD where I have been for about the last 50 years, and that has been a bit of an adventure having been based in Melbourne, the Gold Coast, Brisbane, Townsville, Nowra, Canberra, and a few years in South America in Chile. That has equated to about 26 houses or thereabouts, and, yes, my wife is still with me.

Hardship posting—while based in Santiago, Chile—skiing at Portillo.

I have reflected on the engineering that I have been exposed to over more than 50 years of practice, the broad coverage that was included in the course at TasUni back in the late 1960s and in particular the opportunity we had from course work in electrical and mechanical apart from our core civil engineering. This has provided a valuable understanding of the related areas of engineering which are invariably incorporated in, for example, any civil engineering project.

My interest in hydraulics and the water industry was certainly sparked through my experience at uni, and I am sure this has encouraged me to maintain my technical skills in these areas, notwithstanding a number of office management roles. How could I forget doing water hammer analysis by hand? I retired from full-time work about 5 years ago but seem to be continually involved in all sorts of interesting water-related projects, presently working on submarine intake and outfall pipelines—mostly overseas and major water supply pipelines here in Australia as well as overseas. Most of these have generally been done from my home "office" which has certainly been the case over the last 9 months with the COVID "lockdowns."

I really believe that the broad coverage of the Engineering Degree at UTAS with the opportunity to specialize during the 3 months of honors provided a great basis for my professional engineering activities, and yes I am still registered to practice these pursuits.

Maybe a failing is that I still don't really have an end plan to full retirement.

Roger Stilwell (1962). BE(Hons), MScEng (Southampton, UK), Bachelor of Engineering 1962

I am an alumnus of the University of Tasmania having graduated in 1962 in the faculty of engineering with honors. I was born in Sandy Bay and lived in St Helens during the war and thereafter on small farms in the Deloraine area. My father died during the war, and my mother married again; I was one of eight children.

I began university in 1957 when campus began moving to Sandy Bay from the Domain. I attended the Sandy Bay campus engineering school and graduated in 1961 but stayed on to complete an honors assignment over the summer and was awarded my degree in the autumn of 1962.

Final year subjects were structural engineering, hydraulic engineering, and electrical (power) engineering. I lived variously in Sandy Bay, Hytten Hall (briefly), and Ferntree. I was in the university amateur football team and was selected to represent Tasmania three times and the Australian Amateur side once.

I was awarded a scholarship to attend University of Southampton, in the UK, in 1962 where I began postgraduate study in the subject analysis of shell structures (Arch Dams) using computers. I was awarded an MSc degree in 1964. Subsequently, I worked in London over the winter, doing structural design for the University of Leeds development project.

On return to Tasmania in 1966, I worked for consulting engineers in Fowler England and Newton on the design of sewer systems connecting Kingston to Sandy Bay and then to the Hobart treatment facility. In 1967 I moved to Sydney and worked with Ove Arup doing structural design on components of the Sydney opera house and then the design of microwave link towers connecting Adelaide to Perth. I moved to Canada in 1967.

In Canada, my work has been varied: nonstandard design of a highway bridge, a 12-month assignment to Ecuador for identification, and prefeasibility evaluation of five small hydro projects on the western slopes of the Andes mountains; I took an Engineering Economics postgrad course at the University of British Columbia. Back in Canada I worked on the following: a 12-month design assignment for the University of British Columbia TRIUMF high-energy particle accelerator, structural engineering assignments and feasibility study assignments on hydroelectric projects, and a structural design project in Saskatchewan on an electric arc furnace building for processing scrap steel.

I had two assignments to Indonesia over 3 years; in Lombok Island and in Central Java I worked on Irrigation Water Resources Development projects training young engineers in design and economic assessments of irrigation projects. In Canada, I was project manager for a 3-year municipal upgrade project for 5000 population town in Northern British Columbia including water supply and treatment, sewage collection and treatment, public swimming pool, and curling rink. I have been on hydroelectric project design assessment assignments to Brazil, Peru, Thailand, Vietnam, and the USA. On retirement, I worked for 5 years with a British Columbia Hydro design team on large hydroelectric units in BC and as special reviewer on several large project legal cases.

Lately, I have assisted my wife who began a school for Syrian refugee children 10 years ago, enrollment 400, in a desert town in Jordan where adult refugees are refused permission to work in most jobs.

David Swifte (1964)

Pre-graduation

I started my engineering course when the university was in the early stages of moving from the city down to Sandy Bay. I had morning lectures starting at 8:45 am in Sandy Bay and one lecture and drafting classes in the city in late morning or afternoon. I travelled by train from Claremont into the Hobart central station and rode a bicycle down to Sandy Bay. If late for the 8:45 maths lecture, I was locked out. By the time I graduated, all lectures were at Sandy Bay.

The various engineering disciplines were all covered for the first 3 years with discipline specialization in the fourth year. This was beneficial to me because my strongest subject related to power generation and distribution and I intended to specialize in that, but after vacation work in a Sydney manufacturing facility, I changed to civil engineering and am glad I did. Information technology was in its infancy, and I had difficulty grasping the concepts and detail. Survey calculations were done manually as calculators did not exist. A logbook and slide rule were useful tools to do any checking. Drafting was done manually with a board and tee square, which I still have. For the first 2 years, drafting was at the Hobart Technical College.

The professor of engineering was Arch Oliver. I can visualize most of the other lecturers, but the only name I can remember is Mac (Malcolm) Gregory. Professor Oliver was very informative when we were having workshop sessions in the new building.

Post-graduation

In 1964 I was employed by Gutteridge, Haskins, and Davey. They had just been engaged by the Metropolitan Water Board to investigate and develop the Risdon Brook Dam. During the feasibility stage, I calculated the quantities for the various alternatives; then I was involved during detail design and was the company's field engineer during construction. They subsequently transferred me to Melbourne where I was involved in four bridges along the Tullamarine Freeway.

In 1969 a contractor from the Risdon Brook Dam convinced me to become their construction manager for development of plant services at Gove in the Northern Territory. This involved management of all trades. I then went to W.A. where I worked intermittently for a total of 20 years. I also undertook developments in Queensland and N.S.W. During my W.A. period, I had 7 years managing the minor construction and maintenance of Hamersley Iron facilities in Dampier, Tom Price, and Paraburdoo. Short-term closures of sections of the plant were often necessary so that local maintenance could be performed, and there were biannual total shutdowns of the whole plant when major changes and improvements were undertaken. We had to employ up to 450 short-term personnel and work 12 h shifts × 7 day weeks. The planning, procurement, arranging temporary accommodation and equipment, etc., involved a separate team for months prior.

My employer for 35 years was Fluor, the Australian subsidiary of an American company. In the 1980s, they changed from just doing construction to total project management. Between projects, I was involved in their estimating division. The majority of my projects were development of mining facilities in isolated areas, starting with feasibility studies, and going through to commissioning and handover to the operators. The projects involved the crushing and processing of ore (iron, copper, or gold), plus the associated stockpiling, train-loading, transportation, and ship-loading systems, i.e., roads, bridges, railways, and port works. Most projects required the construction of accommodation and infrastructure facilities, both temporary and permanent. It was the management of a total project where the university multidisciplinary training was welcome as I could understand the themes that the project discipline leaders were espousing.

The Ernest Henry Mine (a new copper mine development near Cloncurry in central Queensland) and the Mining Area C to Yandi Railroad (part of a new iron ore mine development by BHP Billiton in the W.A. Pilbara) won IEA and a number of industry awards in 1998 and 2003, respectively.

Observations

- Civil engineers have a broader approach to an issue than the other disciplines. This is possibly because civil parameters are broader.
- A high proportion of the other disciplines stay in their specialty, but those that do broaden out are very good analytically.
- Few Australians understand what engineers do to achieve the developments that politicians take credit for. More public education of what is involved is required, possibly starting with parents of schoolchildren.
- As with all professions, the nature of engineering practices is constantly changing, particularly the impact of emerging technologies. Artificial intelligence, digital disruption, and globalization are some examples.
- While technical skills are very important, engineers must also accommodate broader skills such as entrepreneurialism, business acumen, and social impact.

- Project management requires a basic understanding of all engineering disciplines, plus communication skills, risk management, and quality assurance. It also requires some understanding of the workings by other professions, e.g., the legal process.
- Project success relates to cost, time, and safety. It involves teamwork where no single component can dominate.
- Successful project management requires constant interaction with the client and with the discipline leaders within the project team and local authorities.
- During the design process, regular reviews by the construction manager can help identify any issues that would cause delays and extra costs during the construction phase. Typical examples are clashes between the design location of piping or cable trays and the structural or other aspects of the building. Also, construction input into the format of specific purchase orders can ensure the items get delivered in the required sequence and timeframe.

David Tanner (1967)

On graduation, I sought the advice of Professor Archie Oliver who insisted that I needed to consolidate my design knowledge before venturing into construction work, so I took up a position with Laurie & Montgomerie Consulting Engineers in Sydney in January 1968.

After suffering several months of simple design work on roads and drainage for various PNG projects, I was keen to help build them, but the company refused to send married men to the field, even though Alex was looking forward to teaching there. We then decided to take a job with Mt. Newman Mining in Port Hedland and spent the next 2 years working on the biggest mining project in Australia at that time. Working alongside the large American company of Bechtel Corporation was wonderful experience and opened up many future opportunities for me across the world.

At the end of 1970, we decided to migrate to Canada and eventually wound up in Vancouver, where I landed a job with a contractor on the Similkameen Copper Mining project near Princeton BC, again working with Bechtel Corporation. We spent a good deal of time during that winter learning to ski and the next year took up an offer of a design job in Bechtel's Montreal office, working on the Sept-Iles Iron Ore project and the new Greenex lead-zinc mining project in Greenland.

When the construction crew from BC came through Montreal and discovered me working on the design of their next project, I suddenly found myself co-opted onto the team headed for Greenland. I spent the next 18 months 500 km north of the Arctic Circle at Maarmorilik on the west coast of Greenland, working with an international crew of American engineers, Canadian miners, Danish construction guys, and Swiss mountaineers—an unforgettable experience! Alex worked at various jobs around Europe, and we caught up in Copenhagen on my R&R breaks every 3 months, travelling around Britain, skiing in Switzerland, or relaxing in Spain.

Bechtel then offered me a job in Brazil, but we decided to take some time off and settled in Dartmouth, England, where I rebuilt an old gaff cutter, Ayesha. After heading up to Edinburgh to work with a precast concrete manufacturer and then building bridges in northern Scotland, we returned to Dartmouth to sail Ayesha across the English Channel, around Brittany, and into the French Canals at Bordeaux. We spent the next year cruising the Mediterranean Sea, from Corsica to Sardinia, and then across to Italy and island-hopping through Greece from Corfu in the Ionian Sea to visit many of the Aegean islands. In 1976 we decided it was time for a visit home and sailed Ayesha across to Malta where we left her, expecting to return later for the return leg to Britain. We never did, having found ourselves heading for parenthood and another change in direction on our return to Tasmania.

I then spent the next 2 years setting up a precast concrete business in Burnie, Townsville, and Hobart, producing L-shaped wall panels for domestic and commercial construction. In 1980 I was offered a position with Hydro Tasmania, supervising construction of Bastyan Power Tunnel and Power Station at Tullah, and so began 9 years of work on the West Coast of Tasmania building hydropower schemes.

In 1983 we moved to Queenstown to work on the Gordon-below-Franklin scheme, which was shut down by the Hawke Government on first of July. We then moved onto the King and Henty-Anthony schemes and were able to influence much of the design work from the field. After building Henty Dam and starting White Spur Dam, I took over as engineer in charge at Crotty Dam, including work on the 7 km headrace tunnel and John Butters Power Station.

With a growing family (Zoe, Andrew, and Mae), we decided it was time to return to Hobart in 1988, and I accepted a position with the DMR as bridge construction engineer for Southern Tasmania. When the construction workforce was restructured into Works Tasmania and corporatized for sale to the private sector, I decided it was time to leave and set up my own business in 1996. After working on various road and pipeline projects across Tasmania, I spent some time as drilling engineer and exploration manager for Great South Land Minerals in the search for oil and gas onshore Tasmania and later building a mineral processing facility for Zeehan Zinc.

In 1998 I was asked by Hydro Tasmania to act as contract superintendent on the repair and upgrade of the 10 km headrace and tailrace tunnels for the Poatina Power Scheme. Since that time I have worked with Hydro Tas as a consultant on nearly every large civil engineering construction project over the last 20 years, including the Poatina Reregulation scheme, dam raising, and spillway upgrades at Cethana, Devils Gate, and Paloona Dams on the Forth River scheme, Trevallyn Tunnel repairs and repainting, downstream filter buttresses at Lake Echo Dam, Dee Lagoon Dam, and Lake Binney Dam, reanchoring and strengthening of Catagunya Dam using 92 of the world's largest posttensioned anchors, Rowallan Dam spillway wall replacement and embankment reconstruction, and the recent upgrade of Murchison Dam in addition to a number of smaller projects during that time.

In 2014 I set up Tas Renewable Energy P/L and negotiated contracts with Hydro Tasmania for sale of water from the diversion of catchments into existing storages at Lake Newton (Tyndall Creek), Lake Plimsoll (Shaded Creek), and Henty Canal (Stitt River). We are also in the process of developing a small hydro scheme at Lake Cumberland near Trial Harbour, with further stages to follow.

Over the last few years, I have also been engaged on a number of other major dam projects, including the planned safety upgrade and anchoring of Burdekin Falls Dam in Queensland (working with Stantec Consulting) and the design and construction review for Mill Creek Dam on the Duck River Irrigation Project in NW Tasmania.

I have promised Alex to seriously consider retiring by my 75th birthday next year, so we'll see how that plan works out!

For my sins, I have recently been appointed as an engineering advisor to the Resource Management and Planning Appeals Tribunal (RMPAT) for a period of 5 years.

I have also enjoyed being part of the UTAS student mentoring program over the last 6 years, giving us the opportunity to help guide the next generation of engineers through their final years of study and selecting suitable employment paths. I commend this program to all experienced engineers as a means of giving back something to our great profession. Unfortunately, the program has presently been suspended due to the COVID virus but will hopefully be up and running again in 2021.

On the recreational front, we have continued sailing as a family over the last 33 years in a number of different yachts, including the circumnavigation of Tasmania in 1995 in our small ferro double-ender "Stormalong," cruising to New Zealand, the Pacific Islands, and East Coast of Australia in our Herreshoff Mobjack "Star" from 2008 to 2015, and recently becoming the proud owner of "Heritage," a Herreshoff Bounty 65 ketch with plans for some extended cruising in local waters.

Chanh Tran (1965)

In the Far East, people remain humble and usually so does this editor/speaker of Asian descent who has a lifetime worth of experience in the engineering profession working in the private sector.

However, I'd like to share with you some points I have noted throughout my 50 years with the industries. My name is Chanh B. Tran. Julie and I have been living in Texas, USA, since 1980. We're both retired now and just enjoyed a well-deserved jubilee tour of VANZA, a well-known acronym for Vietnam, Australia, and New Zealand. The best part of it was a self-guided half day visit of Hobart CBD and Sandy Bay campus of the University of Tasmania.

Oh yes! Such a lovely place my alma mater with lots of memories!

As it happened, I often simplify events to save space and time. On education, I learned a lot from my dear mother and father at home together with five other siblings. The parents were eighth graders, and the siblings all college degreed high school teachers in Vietnam, the USA, and Canada by the way.

As for me, Vietnamese public primary school, French lycee in Saigon, and UTAS, Australia, were all excellent, and I always honor all my teachers at all levels as I remember most names and their devotion to educate us their pupils or students.

As an electrical engineer, I have worked for a milk plant in Vietnam for 11 years, a superminicomputer manufacturer in FL for 12 years, and a global computer giant for 10 years—all of them on the Fortune 500 list.

The training by UTAS has helped me most in my first job as plant engineer supporting the food production dealing with construction, expansion, installation, operation, maintenance, etc. The course business practice for engineers in my opinion could be expanded to cover industrial management and inventory control, but I gained enough experience to set up useful systems on the job. We had no major accident in the plant with a total of 300–500 workers at the peak activity!

In the USA, I took more courses or self-taught many topics in computer science and software engineering like architecture, networks, and programming to hold the positions I got directly or via contracts with technical staffing agencies. As you may have observed, the industry makes giant leap and products went through updates or became obsolete altogether. More packages to learn, more debugging!

Well, I am glad I have retired; the burden is on the younger generation now, and they will advance far!

For a long time and still now I think AI and genetic engineering will be dominant in the landscape in the next 20–30 years.

To our grandchildren, I would say: Study what you like best to a level comfortable to your situation, but also make an effort to complete a college degree that will certainly help with your first job!

5.4 Common Theme

The common theme of the UTAS engineering graduates from the early days of the UTAS Faculty of Engineering is the benefit they gained from the broadly based curriculum which had 3 years in common for all students. Graduates went on to stellar careers ranging from economics (Kevin Cox), telecommunications (Gregory Crew), and civil construction (Richard Green). Many were employed by the Hydro-Electric Commission in Tasmania at a time when major dam and power station construction was underway.

Besides the broad nature of the curriculum, the relatively small size of the faculty and the expertise and dedication of the lecturing staff, particularly Professor Oliver, are common themes of the graduate's recollections.

With the advent of microelectronics and computers in the early 1960s, the course needed to change to accommodate these new fields. The 3-year common course was reduced to 2 years and is now only 1 year common to all engineering undergraduates. How this transpired and the internal politics of curriculum change has been covered in detail in Chap. 4.

5.5 1970s Graduates

The 1970s saw the emergence of graduates who chose to be specialist electrical, mechanical, or civil engineers. Only civil engineers studied geology. Mechanical engineers could study advanced structures and stress analysis but were more focused on thermodynamics, fluid mechanics, and mechanics of machines. The study of control was jointly embraced by mechanical and electrical students. Signals, analogue electronics, and power systems were firmly in the domain of the electricals. Engineering computing emerged as a common discipline with the first general-purpose computer in Tasmania, an Eliot 503, jointly purchased by the university and the Hydro-Electric Commission in 1964.

Stuart Francis Barwick (1975). Bachelor of Engineering 1975

Perhaps the most enduring thing from the faculty days for me was advice from Archie (Oliver) to his students to "make a start." This was about solving differential equations in hydraulic analysis via iteration. We get into a tutorial, and we just wouldn't have a clue what he'd been talking about. There would be this impossible problem that he put in front of us. "Make a start" is very good advice which has got a wide application. I've quoted it many times in my career to lots of people. Another one from several lecturers who dealt with calculus was "Infinity is often not very far away." Again, for me that has a wider meeting which stemmed from the fact that UTAS in the 1960s had a number of individuals and teams who were world leaders in their research. I mean some of them had come from World War II manufacturing and stuff like that and you know we are still practicing at the university. So, they were world leaders and researchers in their fields and not just in the engineering faculty. We were also exposed to physicists and chemists—Tom mentioned Harry Bloom (what a character he was, wasn't he?)—and the geologists; you know all of these people were world-class. So, the idea that Tasmanians were not shackled by location or population size has endured for me, and indeed the broad grounding which was deliberately instilled in UTAS's engineers opened doors for us nationally and internationally. You know I think both the previous speakers have alluded to that.

Personally speaking, when I look back on events which started for me in 1969, the keyword is gratitude. It must have been frustrating for my lecturers and examiners to see students like me who are not realizing our potential academically. But they didn't give up on us, and eventually for me at least the message sank in that I would have to do some hard yards if I was to graduate. It wasn't just staff who helped me. When I failed my fourth year, there was a senior student whose name appears on tonight's list who gave me some very strong advice and essentially formed my career path for me. The irony was that if I had graduated in 1970 as I was supposed to, there would have been no employment for me. Several of the people who graduated when I should have never worked as engineers. That was just the labor market at the time, but as an undergraduate I was able to get a part-time job as a draftsman while I finished my degree, and suddenly the world of engineering opened up for me. My engineering degree has been my ticket to earn and my ticket to ride, and it sure is a great ride. As a UTAS engineering graduate, I've had the world at my feet, and I still derive a great deal of satisfaction from devising solutions to those impossible situations that face me.

Graeme Booth (1977)

For many of us, our first degree is the one most fondly remembered.

What Do I Remember?

- A cold, somewhat uninspiring building at the bottom end of the Sandy Bay campus
- My first real engineering unit—first year, Dr. Peter Doe as the lecturer
- A long-established reputation of engineering students being big drinkers
- Fascinating laboratories
- The larger-than-life character of Prof. Arch Oliver
- A total absence of females in my "year"
- An eclectic mix of lecturers—admirable and distinguished folk all
- Picture of the construction of Warragamba Dam in the labs
- Someone in first year attracting a lot of attention for buying a Casio calculator
- Fortran programming (or was it Basic), faithfully stored on rolls of punched paper tape
- The mystery of getting high distinctions in all the hydraulics units that Arch Oliver presented when I felt I understood so little
- The "generosity" of my fellow students in finding a parking spot for my mini car on top of the steps at the front of the engineering building when I emerged one day

Thoughts on My Engineering Degree

I reflect very positively on my time studying at the Engineering Faculty at UTAS—particularly of course an outstanding group of lecturers. I also value the start engineering gave me to a fulfilling career. While the last 25 years have been in corporate roles, I found the first 7 years as a bridge designer wonderfully fulfilling. I now work as an executive coach.

So many engineering graduates go on to do diverse and amazing things with their careers. Engineers generally confront challenges with a "can do" attitude and see issues as challenges and opportunities rather than threats. The leaders and staff at UTAS purposely created degrees that developed strong disciplines that fitted us for success in a diversity of work. What a terrific legacy the engineering faculty created over decades!

Ian Malcolm Harris (1978) (Transcript of Video Presentation October 31, 2022)

It's a delight to be here. A big thank you to Peter who took me on a cook's tour of the faculty and the lab late last year. It was wonderful to walk back through and see all the experiments and the bits and pieces and to chat about the wind tunnels, double and triple pendulums, triangular dynamic vibration plate, machine and testing tools, and the lecture rooms and labs. But let me say there's nothing a pitot tube can't do when handled properly, so one of the things I did learn was the value of the pitot tube, and it took me to places I'd never believed such as air systems in paper machines and bulk material handling. When I was quite newly graduated, dilute phase pneumatic conveying systems demanded attention, and measuring velocity pressure accurately and easily with a water manometer hooked onto a pitot tube shoved through a hole drilled into conveying pipe closed my understanding loop. It was just quite astonishing to me. Along the way there were some painful lessons learned too—beginning with the course curriculum which was very intensive. There was a wonderful balance of theory and practice—lots of maths, loads of theory, and plenty of testing to destruction, plus all that vacation experience.

I will say I feel privileged to have had a lovely career; the last 25 years was spent in heavy industry with Aker Kvaerner and Worley operating within oil and gas and mining market sectors, with focus on winning work and also business improvement.

Nothing, however, takes away the memories of Malcolm Gregory who always taught from first principles and kept difficult things simple. I recall to my embarrassment even now, marching in one day during second year after a Material and Structures 1 lecture with a beautiful matrix to a simple static truss problem, or so I thought. He took one look at it, got his pen out, went diagonal-diagonal, cross-cross, and then wrote really slowly and laboriously across the face of paper I'd given him "there is always an easier way." I'll never forget that lesson. It was like, here is a remarkable man. He held not back one bit, but he was a fantastic guy, just as Prof. Oliver was too. And Archie, you know, remains very, very cherished. We were all a little terrified of him, for no good reason. I'm glad you mentioned his recording, Peter (and with it some worthy anecdotes about you too!), because it is really worth listening to, likewise that of Malcolm Gregory. He has a recording too and it's delightful to hear those voices again. It's like sort of hunching around our gramophone and hearing familiar sounds over again; it's just fantastic. So, I'd recommend that if you haven't yet done so.

But all in all the practicality of the training was wonderful though a bit frightening at first. The memory that we started our first year early with drawing (on the board) and Peter's Engineering 1 lectures has stayed with me. Likewise, his justifiable surprise at my struggle with basic cam design. We started in a workshop and learned the practical skills in parallel with the theory. It was an unbeatable combination, and I just hope it stays that strong.

One thing—having worked in many global locations, I can assert that Tasmanian engineering provided the best global exemplars—Comalco, and also ANM, in particular. The teaming under strong leadership was unusually wonderful to behold—all things considered, extraordinary talents abounded—better than anywhere else to my experience, and there were plenty of elsewheres. That says a lot for the island culture and also for all of Prof. Oliver's special team! And a special thank you, Peter.

Neil Mickleborough (1979)

Neil graduated with PhD in structural engineering in 1979. His research at UTAS was centered in the dynamic behavior of reinforced concrete. Research supervision was done by Malcolm Gregory and Frank van der Woude. The external examiner of the thesis was Sergio Giudici. Neil also received a Full Blue in Athletics from the University.

After graduation, Neil worked initially at the PWD as a bridge design engineer and then moved to Sydney working as the Senior Bridge engineer and Chief Bridge engineer with Rankine & Hill (later the company becoming part of Connell Wagner).

After his time in consulting, he joined the University of New South Wales as senior lecturer where he was actively involved in structural engineering research and substantial consulting in bridge engineering throughout Southeast Asia.

In 1991 Neil joined the Hong Kong University of Science and Technology (HKUST) as a founding faculty and had some 10 years in numerous positions including assoc. pro-vice chancellor, assoc. dean, and director of various centers.

In the decade of the early 2000s, he was principal of the Vocational Training Council in HK, dean of Engineering in Thailand, and dean and provost at the Higher Colleges of Technology in the United Arab Emirates (UAE). During the time in the UAE, Neil was involved in some construction aspects of the Burj Khalifa, the world's tallest man-made structure at 828 m tall. At one stage, he had five of his former students from HKUST also working on this structure.

In 2010, Neil returned to HKUST as professor, both lecturing in civil engineering and director of the Centre for Engineering Education Innovation, a position he held until 2017.

Most recently (2017–2019), Neil has headed investigation teams for the Hong Kong Government into prestressed concrete issues with bridges in Hong Kong including the Tuen Mun-Chek Lap Kok link section of the Hong Kong-Zhuhai-Macau structures.

Neil has published five books and nearly 100 technical publications. He is presently adjunct professor in engineering at UTAS.

Ian Shuey (1975)

I've learned over time an engineering degree is an entree to virtually any field of creativity and organizational ability, as it focuses on the most difficult learning process—logic and problem-solving. It enables us to measure and quantify how nature works. Other degrees are memory-based.

Nine years of work experience prior to graduating in mechanical engineering was the catalyst for rapid applied engineering responsibilities. As a young *Telecom mechanical engineer*, the challenges of building a state-of-the-art engineering jobbing and maintenance facility at Lampton Avenue (Derwent Park) for the whole state were very satisfying. Overseeing a works program for 60 (often militant) trades artisans was a baptism of fire.

In 1979 I was offered the position of *regional engineer for Tasmania, with the Commonwealth Department of Science and Technology***.** The focus was on the development of threshold technologies that were emerging from an explosion of microelectronics applications in Silicon Valley. I was able to build a unique knowledge of Tasmania's industrial and scientific capacity, particularly in areas of defense supply, genetic engineering, biomedical engineering, nanotechnologies, precision machining, and robotics.

With such knowledge, in 1984, I joined the newly established *Tasmanian Development Authority* (TDA), the brain child of Premier Robin Gray, for a 3-year term securing defense contracts. TDA was a great door opener for me, and over a period of time talking to entrepreneurial business leaders and understanding their qualities and reasons for success, patterns were starting to emerge in my mind about the very nature of entrepreneurship and the possibility of launching something in my own right. Enter Balloon Magic Pty Ltd!

Thirty-two years later, **Balloon Magic P/L** is still operating out of my lounge room! We jumped out of the blocks with bold ideas that businesses and the community leaders hadn't seen nor contemplated previously. Our secret was to be bigger, bolder, and brighter in designing solutions to people's problems. Convention and corporates were our target. Our creations were published internationally, and before long we'd received top world awards and a string of Australian and local business awards. I was once nominated and short-listed as the Tasmanian Business Achiever of the Year, only to be knocked off by Rob Clifford (International Catamarans). There were a dozen visits to the USA and a training program I designed and conducted in Hollywood. Balloon Magic P/L was a low-risk gamble which turned into an exciting, stimulating, and rewarding lifestyle and meeting lots of creative people throughout the world. The world's largest balloon arch (440 m) was built in Hobart over water in 2 hours—serious planning and number crunching!

Ian Shuey, balloon engineer, aged 70, 2020

Dario Tomat (1976). Dario Tomat CPE, FEA, FAICD, APEC Engineer, RPEQ BE (UTAS), BBus (UTAS), MBA (Deakin/APESMA), Grad Dip Co Dir (UNE)

On completing a BE, Dario joined W.E. Bassett and Partners in Hobart working as a building services engineer. After 3 years, he moved to construction as the project services engineer on the new (for 1979) Launceston General Hospital.

Returning to Hobart in 1982, he worked for a period as a policy officer at the Department of Premier and Cabinet before transferring to the HEC Energy

Management Centre. He then moved to a role as senior engineer in Energy Resources Planning dealing with early alternative energy planning initiatives.

In the late 1980s, he was appointed contract engineer at HEC Enterprises Corporation working on a range of projects as part of the early formal external consulting undertakings by the HEC.

In 1990 his career moved into manufacturing as general manager of the Centre for Precision Technology. During this time, he renewed his association with the university collaborating on establishing a Bachelor of Technology award in Manufacturing Engineering.

In 1996 with a range of other partners, he established Whetstone Pty Ltd, a consulting group focusing on a variety of project management assignments in the forestry industry and in supporting defense.

Dario was a member of the Royal Australian Navy Reserve as a marine engineering officer retiring with the rank of commander.

He had a long association with the Association of Professional Engineers serving in a number of honorary officer roles nationally including several years as national president. He then became the chairman of the Assessment Panel for registration of professional engineers through Professionals Australia.

In 2015 he was appointed as a director to the TasPorts Board. He is a former chairman of the National Trust in Tasmania and chairman of the Brain Injury Association of Tasmania.

Dario is married to Christine and has three children, who all have graduated from the University of Tasmania.

5.6 Endnote

This chapter has presented the undergraduate experiences of a sample of students who studied at the University of Tasmania in the latter half of the last century and were motivated to participate in the faculty's centenary celebration. Several of the UTAS Rhodes Scholars contributed at the author's request. Their careers as professional engineers varied widely. Not many were still engaged with engineering design or at the engineering "coal face" up until the time they retired—they had become managers, company directors, and, in many cases, mentors. Many are still working as consultants. It is hoped the readers of this book, who may be young engineers, will be inspired by their stories. Engineering education in the future, as presented in this book's penultimate chapter, may not be very different from that experienced by the contributors to this chapter. The basics of science, mathematics, and technology will be delivered; there will be competent and inspiring lecturing staff to guide them; they will learn through their "hands-on" experiences in the laboratories and workshops; and most importantly they will benefit from working with their peers from diverse cultural and ethnic backgrounds.

Chapter 6
Recruiting International Students

Abstract This chapter looks at the period of internationalization in engineering education which saw Australian and New Zealand universities actively recruiting international students and universities sending staff to neighboring countries in the Asia-Pacific region to deliver engineering courses and set up campuses. Internationalization started with the Colombo Plan which provided scholarships for students to study at Australian and New Zealand universities. The chapter details the University of Tasmania's experience with students coming from Indonesia and then Vietnam in the late 1960s. Then followed recruitment visits from academics and administrators to SE Asia which resulted in Memoranda of Understanding through which UTAS admitted graduates from Malaysian and Singaporean colleges with various degrees of advanced standing according to their college record. The next and continuing initiatives were 2 + 2 programs with universities in China in which students studied a mix of local units and units taught by visiting UTAS academics. There was also a scheme for diplomates from the Australian College of Kuwait to transfer into the UTAS BEng program with up to 2 years advanced standing.

6.1 Introduction

This chapter chronicles the author's and others' personal experiences with recruiting international students from the Asia-Pacific region since the early 1960s. With the advent of the Colombo Plan (see below), undergraduate and postgraduate students from the Asia-Pacific region began to arrive in Australia to study. Whereas previously the student body was principally local (Australian) students, there was a significant increase in international student numbers. The photo of final year University of Tasmania students in 1961 (Fig. 6.1) contrasts with that of the following year (Fig. 6.2) which includes international students.

P. E. Doe, *Prospects for Professional Engineering Education in the Asia-Pacific Region*, Education in the Asia-Pacific Region: Issues, Concerns and Prospects 74,
https://doi.org/10.1007/978-981-96-3197-1_6

Fig. 6.1 Engineering students in 1960. Front: Alan Shaw, Sandy Buchan (deceased), Ian Baird, Peter Doe, and Greg Walker (deceased). Back: David Mitchell, Richard (Dick) Groom, Hugh Murphy, Ted Pitman, Robert Campbell, and Brian Cousins (deceased)

As detailed below, the University of Tasmania successfully established a pathway for students from Vietnam to enter the UTAS BEng program with advanced standing. Then followed students from Malaysia and Singapore. Attracting students to study in Australia was very competitive both between universities in Australia and universities in the UK, USA, and Canada. Education fairs were held in the major SE Asia cities with delegations manned by university academics, agents, and support staff who set up "stalls," provided handouts, and fielded questions from prospective students and their families.

Some Australian, UK, US, and Canadian universities established campuses offshore or had negotiated agreements with in-country higher education institutions to teach into their courses and award internationally badged degrees.

The peak of international student enrolment occurred just before COVID in 2019 with the majority of students coming from China.

This chapter recounts the international student recruiting experience of the University of Tasmania in Sarawak, Sabah, Hong Kong, Penang, Kuala Lumpur, Singapore, Indonesia, Brunei Darussalam, and Vietnam in the late 1990s. The chapter then details the UTAS experience with students coming from Indonesia, Vietnam, Malaysia, and China; the more recent experience of UTAS staff teaching in Kuwait and China; and the disruption from the COVID-19 pandemic.

Fig. 6.2 UTAS School of Engineering final year students 1965 with Koes and Tris

6.2 The Colombo Plan

The Colombo Plan was a postcolonial initiative launched in Colombo, Sri Lanka, in 1951, initially comprising Commonwealth countries Australia, India, New Zealand, Sri Lanka, Pakistan, Canada, and Britain. The Colombo Plan has now expanded to include the following: Afghanistan, Bangladesh, Bhutan, Brunei Darussalam, Chile, Fiji, Indonesia, Iran, Japan, Korea, Laos, Malaysia, Maldives, Mongolia, Myanmar, Nepal, Papua New Guinea, Saudi Arabia, Singapore, Thailand, USA, and Vietnam.

Supported by the Colombo Plan, students from several Southeast Asian countries chose to study at the University of Tasmania. Two students from Indonesia were the first female students to enroll in engineering: Koesnowarso Koesmarihati (Koes) and Trismiati Harsono (Tris). Koes became the first female president-director of the Indonesian telecommunications company Telkomsel. Another student from Indonesia, Jonathan Parapak, married a Tasmanian and rose through the ranks of the Indonesian public service to become secretary-general of the Department of Tourism, Posts and Telecommunications.

Following is the citation for Koes, as we knew her as an undergraduate, on the occasion of her being named as the 2023 Australia Alumni of the Year and receiving the Distinguished Alumni Award and Doctor Honoris Causa in Engineering from the University of Tasmania:

Koesmarihati Koesnowarso and Her Trailblazing Journey to Crack the Telco Industry's Glass Ceiling

The 2023 Alumni of The Year, Koesmarihati, graduated with a bachelor's degree in electrical engineering from the University of Tasmania in 1965 under the Colombo Plan scholarship scheme.

The telecommunication industry in Indonesia remains a male-dominated sector that rarely puts women in C-suite roles. There is one extraordinary woman, however, who has made a permanent crack in the glass ceiling. Her name is Koesmarihati Koesnowarso, and she is recognised as the most influential woman in Indonesian telecommunications history.

She had a long career with state-owned telco PT Telkom and later its subsidiary Telkomsel, creating a series of regulatory and technological breakthroughs. As the first CEO of cellular phone operator Telkomsel, she has been credited for the company's success in becoming the largest player in the industry with wide coverage across the archipelago—the position the company has attained until now.

When she first took office in 1995, Telkomsel made a breakthrough by separating telecommunication services and mobile devices, making the devices much more affordable for the market, and reducing the average cost of a device from Rp14 million to Rp1-4 million. The company first carried out this breakthrough independently before finally being supported by the government on a national scale in 2000.

"I said, the company already had a huge network; let's just focus on selling the network through phone cards. For the device, leave it to the market. Let's introduce an open distribution channel, establish dealers all over the country, and create more jobs," said Koesmarihati on the then-controversial decision in a recent interview at her home in South Jakarta.

Following her retirement from the company, she was appointed as an official at the Indonesian Telecommunications Regulatory Body (BRTI), representing the public. She did not oppose the liberalisation of the country's telco sector, but she insisted that local players must also have a share in the market.

At BRTI, she drafted a road map for cellular phone number portability, which allowed a single code in Indonesia, enabling phone users to change operators without having to lose their familiar phone number.

Now 81 years old, Koesmarihati remains an active member of the Indonesian Telecommunications Society. She also teaches at Telkom University and the University of Indonesia and serves as a consultant, board member, and advisor of the Telecommunication and Information Accessibility Body (Bakti).

Brave being the operative word

Koesmarihati graduated with a bachelor's degree in electrical engineering from the University of Tasmania in 1965 under the Colombo Plan scholarship scheme. Along with fellow Indonesian student Trismiati Harsono, they became the first women to graduate from the university's School of Engineering.

"We were called [the] two brave ladies who took up engineering in our faculty," she said, adding that Trismiati now resides in the United States.

Koesmarihati attributed their bravery and courage to her parents, Koesnowarso and Koesmarinah. Serving as a government forestry inspector in West Java back in the 1940s, Koesnowarso once sat down the family of seven children, saying that he had to sign a statement siding with the Dutch colonial government.

"He said that he would not be able to face us if he became a chameleon (country's traitor). He said the Dutch colonial [officials] would target him so that he would be on the run for a while," Koesmarihati said, reminiscing about how the colonial officials were shooting at her father's car.

Her father came back after a couple of years in hiding but passed away in 1951, leaving her mother to work hard to raise the children and make ends meet as they relocated to her grandmother's house in Madiun, East Java.

Back then, Koesmarihati was already a bright student with a talent for math and science, but the family could not afford to send her to college. Fortunately, she was selected for the Colombo Plan as she was one of Madiun's best-performing high school graduates.

Telecommunication the path to gender equality

Following her graduation from the University of Tasmania, Koesmarihati worked briefly at the Hydro Electric Commission in Hobart and PMG (now Telstra) in Melbourne before finally deciding to return home to work in the infrastructure and telecommunications industries.

In Indonesia, she worked at the state utility company PLN's Laboratory and for the Dutch company Nederlandse Kabelfabriek (NKF). Koesmarihati then joined Telkom in 1976, working on a project connecting the network cables in Jakarta and Greater Jakarta. She initiated the utilisation of the latest fibre optic technology, first used in the United States in 1977. The project pioneered the use of fibre optic cables in Indonesia, which earned her the prestigious Satya Lencana Pembangunan award from the Indonesian government for her extraordinary contribution to the country.

In 1993, Koesmarihati was appointed as Telkom's Development Director, in charge of laying the foundation of cellular technology and developing GSM technology. She succeeded in both and was promoted to become the first President Director of Telkomsel in 1995—the first woman in the country ever to lead a telecommunications company.

She said women are certainly as capable as men in any sector. Still, in the reality where gender equality has yet to be achieved, it is important to have male allies to set a level playing field and make sure that women get a seat at the table.

"I don't think I would have been able to rise to the top without the trust of the Minister of Tourism, Postal, and Telecommunication at that time, the late Soesilo Soedarman. There were many contenders back then to become directors, but he nominated me because he said he liked how I never minced my words," Koesmarihati said, laughing.

She said she is glad that it is now a different era where more women are promoted to the top levels faster. However, the telecommunications industry still lags behind as no woman has been appointed as a regional head.

"I hope women's access to technology and their role in the telecommunication sector can be improved. The internet can really expand the horizon of women," said Koesmarihati, who is a recipient of the Distinguished Alumni Award and Doctor of Honoris Causa in Engineering from the University of Tasmania. (Australia Awards 2023)

A photo of final year engineering students taken in 1965 shows Koes and Tris with local and other international students (Fig. 6.2).

6.3 Vietnam

Professor Thong Nguyen was appointed to the chair of electrical engineering in 1974. He was one of the last refugees to escape from Vietnam when that war ended and arrived in Auckland with little or no English to complete an engineering degree. The appointment of Professor Tong Nguyen saw the University of Tasmania develop close ties with Vietnamese universities and an influx of students from that nation. As with the Indonesian graduates, those from Vietnam became successful engineers, managers, and academics.

A photo taken in 1963 of the residents of Hytten Hall on the Sandy Bay campus (Fig. 6.3) identifies eight Vietnamese-born engineering students in the top row: Nguyen Kim Long, Duong Hong Phi, Pham Do Tuan, Nguyen Van Phu, Nguyen Van Dac, Tran B Chanh, Le Q Tam, and Ta Trung Quang.

Fig. 6.3 Engineering students at the UTAS College of Residence, Hytten Hall, in 1963

6.4 Malaysia and Singapore

In the 1980s, UTAS had "twinning" arrangements with tertiary colleges in Malaysia and Singapore. Students graduating from those colleges were admitted directly into the third year of the UTAS Bachelor of Engineering. For a short time, students with engineering degrees from colleges and universities in SE Asia were given 3 years advanced standing. This was ceased when a UTAS civil engineering graduate was demoted from engineer to draftsman at the Hydro-Electric Commission when found to have made a critical error in specifying reinforcing steel in beams he designed. The error was most likely due to a misunderstanding of instructions from his supervision because of a lack of technical English. Problems with technical English fluency resurfaced many years later with Chinese students transferring to UTAS.

The School of Engineering actively recruited students from tertiary colleges in Malaysia and Singapore in the 1990s. The university sponsored visits from three or four academics and a staff member from the UTAS International Office to Singapore, Malaysia, Sarawak, Sabah, and Brunei to visit colleges, attend "education fairs," and hold information meetings. The delegations were always warmly welcomed with considerable interest in Tasmania, Hobart, and the university. I recall a particular meeting in Sarawak seeing a woman weeping at the back of the audience. She was from Tasmania and was with her Malaysian husband, a doctor, who was working as a GP in the remote communities. She was homesick seeing the photos we showed, but her sadness was for the plight of her children who could not be issued with visas for her to take them back to Tasmania without her husband.

6.4.1 Advanced Standing Recognition

As a result of the visits to universities and colleges in the Asia-Pacific region, the Faculty of Engineering entered into Memoranda of Understanding (MOUs) with several Thai, Singaporean, and Malaysian colleges including King Mongkut's University of Technology (Thonburi, Thailand), Srinakharinwirot University (Bangkok), Singapore Polytechnic, Prime College, and Nilai College (Malaysia), and INTI College (Kuala Lumpur and Kuching). The Faculty of Engineering agreed to admit graduates from these colleges into our 4-year Bachelor of Engineering program with advanced standing on the basis of the "percentile system" as an overall guide shown in Table 6.1.

Table 6.1 Advanced standing for Malaysian and Singaporean diplomates

Advanced standing granted	Overall diploma course average	Maths average over 3 units
2 years	65%	65%
1.5 years	55%	65%
1 year	50%	55%

Table 6.2 Assigned mark for grade ranges

Grade	Range	Assigned mark
A	80–100	90
B	70–79	75
C	60–69	65
D	50–59	55
E	40–49	45
F	0–40	20

Source: Memorandum from Professor Frank Bullen, Head of the School of Engineering, to Mark Bennett Executive Offices, Faculty of Science and Engineering, March 13, 2001

Since individual unit marks were often unavailable, the midpoint of grade ranges was used to determine averages and the amount of advanced standing. An example for Singapore Polytechnic is shown in Table 6.2.

6.5 China

In June 2010, the University of Tasmania (UTAS) signed an agreement with the Shandong University of Science and Technology (SDUST), Jinan, China, to cooperate in the area of undergraduate teaching in engineering. The agreement specified, as general principles, that UTAS will provide curriculum and academic support in years two, three, and four of the SDUST Bachelor of Electrical Engineering and Automation degree and provide the opportunity for eligible students to complete their electrical power engineering studies at UTAS in Hobart. Students who successfully completed 2 years at SDUST and then successfully completed a further 2 years of electrical power engineering studies at UTAS were awarded both the UTAS and SDUST Bachelor of Engineering degrees.

The program management identified the responsibilities of both universities. UTAS committed to deliver five units across years two to four in the UTAS telecommunications engineering degree. Each unit was to have a 10-day teaching period. UTAS was also responsible for delivering an English language program at SDUST. For its part, SDUST agreed to pay a fee for UTAS' contribution to curriculum development and teaching into the agreed units. The agreement included a quality assurance program that mandated regular visits of senior faculty to UTAS and SDUST.

The agreement identified a pathway for students to have the option of transferring to UTAS after 2 years of the electrical power engineering studies at SDUST. The offer of enrolment at UTAS was conditional upon the students successfully completing the academic and English requirements for entry. There was also the stipulation that students who did not achieve the required IELTS score for degree studies be required to undertake further English studies at UTAS before commencing their engineering studies. UTAS provided a pre-departure, on-arrival, and ongoing assistance to students transferring to Tasmania.

The agreement between SDUST and UTAS was specific concerning the payment of a service fee to UTAS. It was based on a guaranteed minimum payment of 90 students in each of years one and two and minimum cohorts of 90 in years three and four (less the number of students in each cohort who transferred to UTAS each year).

In March 2011, SDUST and UTAS agreed to deliver joint degrees in telecommunications engineering with the same terms as for power electrical engineering.

This business model was made possible by SDUST gaining permission from the Chinese Ministry of Education to operate an international program for which the students presumably paid a premium. What that meant for the UTAS staff teaching in China was that each class had at least 100 students which was a substantial load for teaching and assessment. Under the terms of this contract, a 72-h UTAS curriculum was delivered in 10 working days. This equated to 8 hours face-to-face contact each day except for one morning and one afternoon session in the 2 weeks of the visit.

A major problem was the mismatch in the degree of specialization between the SDUST and UTAS curricula. The UTAS degree at that time had 1.5 years common to all specializations. Accordingly, the SDUST telecommunications students were required to take a second year engineering mechanics unit. At UTAS the second year students had studied kinematics and dynamics of machines in the first year, whereas the SDUST students only had high school year 12 physics. So, in addition to delivering the engineering mechanics content, the UTAS lecturers had to fill the gaps in the SDUST students' understanding of kinematics and dynamics.

In September 2013, an agreement was signed between the University of Tasmania and the Xi'an University of Science and Technology (XUST). Under this agreement typical of 2 + 2 Chinese Ministry of Education (MOE)-approved programs, UTAS as the foreign university was obliged to deliver teaching into the XUST curriculum. The MOE requirements were that the foreign partner contributes one third of the curriculum content to the Chinese partner which equates to eight 12.5% units over 4 years plus two years of English language proficiency training amounting to 200 hours per year. The first cohort of 19 students transferred from XUST to UTAS in the second semester of 2016.

6.6 Kuwait

In August 2006, I was appointed to the position of head of engineering at the Australian College of Kuwait (ACK), in the suburb of Mishref, Kuwait City. It was established only 2 years previously as one of the first private colleges in Kuwait. It has now university status being renamed "Australian University." While I was there, a new building was built for engineering staff offices, and a hangar was erected around a Boeing 737 aircraft transported from the airfield. When I arrived there were 13 staff members delivering diploma level programs to 323 students in civil, mechanical, and electronics engineering as well as a program of aviation studies overseen by Kangan-Batman College of TAFE, Melbourne, with a pathway leading to qualification as a licensed aircraft maintenance engineer.

The majority of enrolled students were of Kuwaiti nationality, with a mix of males and females. Their educational backgrounds were from government public institutions and British and American private schools. They had little or no experience with a vocational system and were not used to a participative/collaborative approach in learning. Once introduced to the "Australian style" of competency-based training, they responded very well. Arab students generally rely heavily on oral communication rather than reading and writing. They are certainly capable of these tasks; however, it is not a big part of their culture or prior learning experiences. Being a member of an Arabic "clan" is an important aspect of their culture. Students usually worked with their clan and family members and relied on their support. This behavior, unfortunately, continued into formal assessments. Cheating, as understood by Australian staff, in examinations was rife.

Besides delivering classes on mechanical engineering, I facilitated an articulation program with UTAS in which ACK students studied university-level subjects including first year mathematics at ACK so they would have the prerequisites for transfer into the second year of Bachelor of Engineering Technology degree programs at UTAS in Hobart.

My time with the Australian College of Kuwait came to an unexpected end in August 2008 when, soon after returning from annual leave, I was summoned to the principal's office and told that my position had been terminated, my visa cancelled, and that I was required to leave the country within 10 days. The reason I was given at the time was an issue with some students not having completed a unit required for them to graduate on time.

Thinking back on the circumstances surrounding my dismissal, I now feel the reason was more to do with my uncompromising stance on cheating in assessments. I could devote a chapter in this book on the methods and examples of cheating which I encountered—notes on the back of calculators which could only be seen when viewed from a certain angle, students signing with hand gestures the answers to multiple-choice questions, and hearing devices in ears. In one examination, all the students wore the traditional Arab headdress the "Gotra." All but one of the students sat quietly doing nothing for the first 30 min of the exam, while one student wrote furiously, handed in his script, and then left the room. At that stage, the other students started writing with their left hands cupped to their ears.

On another occasion, I was called to the exam room 15 min after the start. The invigilator told me that a student had already submitted his script. Clearly, he had somehow managed to obtain the script, completed it at home beforehand, and smuggled it into the exam room. The staple on his script was not quite in the same position as the others handed out at the time nor was the gray area on the script the same density as on the others. I accused the student of cheating, but he swore adamantly that he had not cheated; he was a fast writer. So, I asked the student if he would be willing to answer the questions again. He agreed and proceeded over the next 30 min to complete the script. It was nearly all correct so, in face of the evidence to the contrary, I had to give him a pass. Several months later, I happened to meet the same student at a football match. In our chat, he did not directly admit to cheating but did explain that in his culture students were used to working in groups. He did not feel confident having to undergo an individual assessment. It made me sad to think that this student, who had clearly demonstrated his competence by writing the exam script under my supervision, was lacking the self-confidence we expect to see in a professional engineer and felt compelled to cheat.

What most likely tipped the scales against me was an incident in which one of my staff accused a student of cheating in an aviation studies examination. The evidence was clear—the student was summoned to a meeting called by the deputy principal to decide what action should be taken. I had strong views about graduating an aviation studies student who was not competent. For some reason which escapes me now, I wanted the student to admit his guilt; I deliberately raised my voice, and he shouted back at me. This resulted in his being graded as incompetent in that unit and not graduating with his peers. I do not know what influence this student or his family had with the owner of the college, but it may have been sufficient for my contract to be terminated.

6.7 COVID-19

The "cash cow" of Chinese students enjoyed by all Australian universities came to a sudden halt early in 2020 when international borders closed because of the COVID-19 pandemic. Not only were Australian academics not able to visit and deliver classes in China, but Chinese students also enrolled in twinning programs were prevented from continuing their studies in Australia. Lockdowns in Australasia started in January 2020. Staff and students were confined to their homes.

When COVID-19 hit at beginning of 2020, I was delivering an "Engineering Design and Build" unit to 89 second year engineering students on-campus in Hobart and 13 students still in China unable to return to Australia. Six weeks into the semester, the campus was closed with no access to the workshop. All delivery was online. How could we deliver a design and build unit with no workshop? The first problem we had was communicating with the students still in China. That problem was quickly resolved by facilitating a high-bandwidth connection for these students, so they had the same online access as local students.

For several years, the "Engineering Design and Build" unit at UTAS had focused on designing and building an entry for the international Warman Design and Build competition. That year's problem was to design and build a robot to navigate autonomously around barriers on a 1.2 × 2.4 m board and deposit different numbers of balls in cylinders in minimum time. Seventeen teams of six students were formed to design a robot. The students were able to produce detailed engineering drawings within the first 6 weeks of the semester, but with the closure of the campus and workshops, a physical build was not possible (except for two teams who had access to their own private workshops).

When the university workshop became unavailable, the "build" component of the unit was changed to "simulate." There was a very steep learning curve for tutors and students to use MATLAB Simscape. The results were encouraging. Students accepted the challenge and were able to program their simulated robots to navigate the Warman competition course with a high degree of realism (Fig. 6.4).

Fig. 6.4 Simulated robot action for "Design and Build" class during COVID-19 campus shutdown

6.8 Endnote

Recruiting international students to Australia from Asia-Pacific countries has matured, over the last 60 or so years from the government funding of scholarships under the Colombo Plan to a system where agents acting for one or several universities have shopfronts in the major Asia-Pacific cities and facilitate student visa applications. Universities are generous with sponsoring tours to agents, teaching staff, and prospective students to their campuses. Faculty staff travel and deliver courses at higher education institutions overseas.

Students who graduated from Australian universities have formed alumni association in many countries. Many Australian-trained international students now have senior positions in their country's universities, and there are many staff from Australian universities who have joined the staff of international universities. But it is not common for Australian students to enroll in overseas universities, even though many of those universities have good, if not better, facilities for learning and research.

Reference

Australia Awards. (2023). *Australia Alumni of the Year; Koesmarihati Koesnowarso.* Retrieved from: https://www.australiaawardsindonesia.org/news/detail/244000322/koesmarihati-koesnowarso-and-her-trailblazing-journey-to-crack-the-telco-industrys-glass-ceiling#:~:text=20%20May%202023-,Koesmarihati%20Koesnowarso%20and%20Her%20Trailblazing%20Journey%20to%20Crack%20the%20Telco,the%20Colombo%20Plan%20scholarship%20scheme

Chapter 7
Delivery Pedagogy

Abstract This chapter is about the pedagogy of engineering education based on the collaboration between this author and Dr. Seeta Jaikaran-Doe. Following a survey of engineering students' engagement with their lecturers' presentations and online material conducted by a final year student, we turned our attention to the technical, pedagogical, and content (TPACK) competency of the UTAS School of Engineering staff. Then followed work on strategies for enhancing student learning in an Australian 2 + 2 Undergraduate Engineering Program and an investigation into the articulation of the 2 + 2 Chinese students into the UTAS engineering degree programs. The chapter concludes with some of the findings of a review of engineering lecturers' delivery practices post-COVID.

7.1 Introduction

This chapter is built around the pedagogical skills and knowledge brought to the School of Engineering with the arrival in Tasmania in 2012 of my Trinidadian wife Seeta Mohammed. She changed her name to Seeta Jaikaran-Doe when we were married and, after a period working as a teacher's assistant in special needs, enrolled in a PhD program in the UTAS' Faculty of Education. From that time until her untimely death in 2021, she took a keen interest in engineering education initiating and coauthoring the journal and conference papers which form the basis of this chapter. Appendix A is included in this book to record and acknowledge Dr. Jaikaran-Doe's significant contribution to the education of schoolchildren in Trinidad, the USA, and Kuwait and undergraduates, postgraduate candidates, and staff in the School of Engineering at the University of Tasmania.

P. E. Doe, *Prospects for Professional Engineering Education in the Asia-Pacific Region*, Education in the Asia-Pacific Region: Issues, Concerns and Prospects 74,
https://doi.org/10.1007/978-981-96-3197-1_7

7.2 Didactic Delivery

Until relatively recently, persons appointed to lectureships at the University of Tasmania (UTAS) and other Australian universities received no teacher training. It was only in 2014 UTAS required new appointees to take a learning and teaching unit. Prior to that time, lecturers copied the style of the teaching they experienced in their undergraduate years which was Confucius-style didactic, "teacher-centered" delivery. The lecturer stood on a dais in front of the class and read from notes. Some lecturers wrote copious notes on the blackboard which the students dutifully copied.

There is something to be said for this teaching technique. If lecturers are "on top of" their subject, they can deliver the content with each of the generally recognized VARK learning styles (Classroom Advancements, 2017):

- *Visual learning*—write notes and draw diagrams on the blackboard/whiteboard; use different colors for emphasis; bring samples for the class to look at and pass around; invite students to write on the board; use body language and gestures.
- *Auditory learning*—speak clearly; speak slowly and use simple words when lecturing to nonnative English speakers; repeat important points; invite students to explain or comment.
- *Read/write learning*—students copy what is written on the board; distribute handouts for students to read; require students to write answers in tests/examinations (not ticking a box).
- *Kinesthetic learning*—use hand tools; make things.

7.3 Information and Communications Technology

From the start of the twenty-first century, communications technology (ICT) opportunities with overhead projectors, and smart whiteboards, appeared in schools. University lecturers were slow to follow.

In 2015 staff members and a final year student in the UTAS School of Engineering surveyed 11 staff and 106 UTAS engineering students. Opinion was sought on the value of online materials and class attendance. Students were critical of lecturers using the traditional "chalk and talk" method but were also not in favor of PowerPoint presentations. Much could be done to improve student engagement including practical "hands-on" activities and more use of audiovisual material. Students perceived lecturers as more research-focused and disinterested in teaching (Turner et al., 2015).

Below are some comments from students:

- "I'd be more eager if lecturers didn't use PowerPoint so much. I personally learn better when I work through an example myself, at the same time as the lecturer is doing it on the board. When PowerPoint is involved, I usually can't keep up."
- "If they show videos or models or other things that give us an idea of how it works, it makes it a lot more interesting."

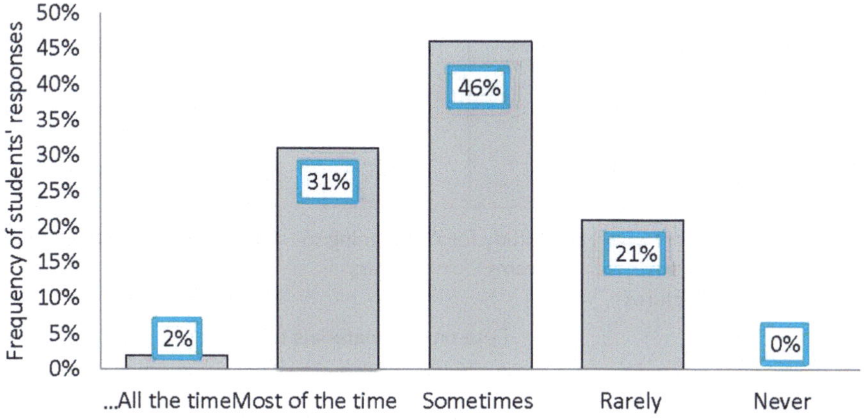

Fig. 7.1 How often did the lecturer make the subject matter interesting

- "Good learning or engagement is to use interactive demonstrations and examples of concepts; visual/examples."
- "I would be more excited and eager to attend class if the lecture did not just consist of someone standing in front reading the slides. It appears almost all engineering lectures are all about the lecturer reading his notes. If they show videos or models or other things that give us an idea of how it works, it makes it a lot more interesting. If I wanted to read some notes about a subject I would use Wikipedia."
- "I would be more excited and eager to attend class if lecturers were better trained as actual teachers which obviously isn't really their actual job."

In this survey, 46% of lecturers keep the subject material interesting and engaging sometimes. Only 2% of students responded that lecturers keep the subject materials interesting and engaging all the time (Fig. 7.1).

Of the students, 95% use the available online materials for supporting what they learned in the classroom or to help them study for exams (Fig. 7.2).

Only 25% of the students showed a preference for face-to-face classes (Fig. 7.3).

It is clear from this study that a lot could be done to improve the engineering students' engagement at this university. There was much criticism of lecturers' unwillingness or inability to adopt modern teaching and learning techniques.

7.4 Technical Pedagogical and Content Knowledge

In order to assess how the UTAS engineering lecturers were making best use of the ICT technologies, a survey of their Technological Pedagogical Content Knowledge (TPACK) was undertaken (Jaikaran-Doe & Doe, 2015).

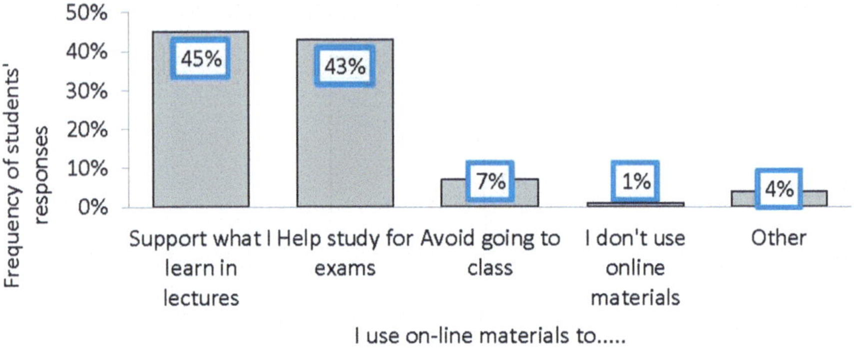

Fig. 7.2 Purpose of online materials

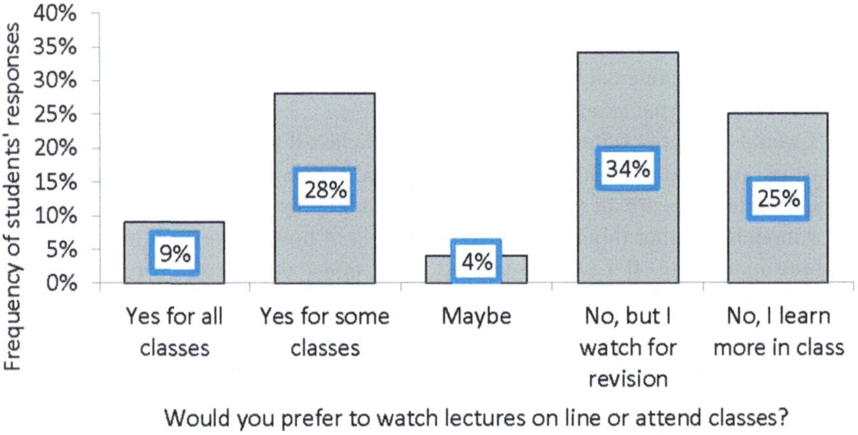

Fig. 7.3 Student preferences for mode of learning

The TPACK framework, originally published by Mishra and Koehler (2006), builds on the work of Shulman (1986) who proposed the integration of two knowledge domains, content knowledge (CK) and pedagogical knowledge (PK), to generate pedagogical content knowledge (PCK). The TPACK framework shown as a Venn diagram in Fig. 7.4 includes a third domain, technological knowledge (TK). The intersection of these domains leads to technological pedagogical knowledge (TPK) and technological content knowledge (TCK).

For their survey, Jaikaran-Doe and Doe (2015) used the "Teaching Teachers for the Future" instrument which was statistically validated to measure the TPACK of preservice teachers in all Australian education institutions delivering education programs (Finger et al., 2013). The survey comprised two sections, each with 24 items. One section is about teachers' confidence to integrate ICT tools in their delivery of content. Each of the questions is prefaced with the phrase: "How confident are you that you have the knowledge, skills and ability to use ICT to do the following...."

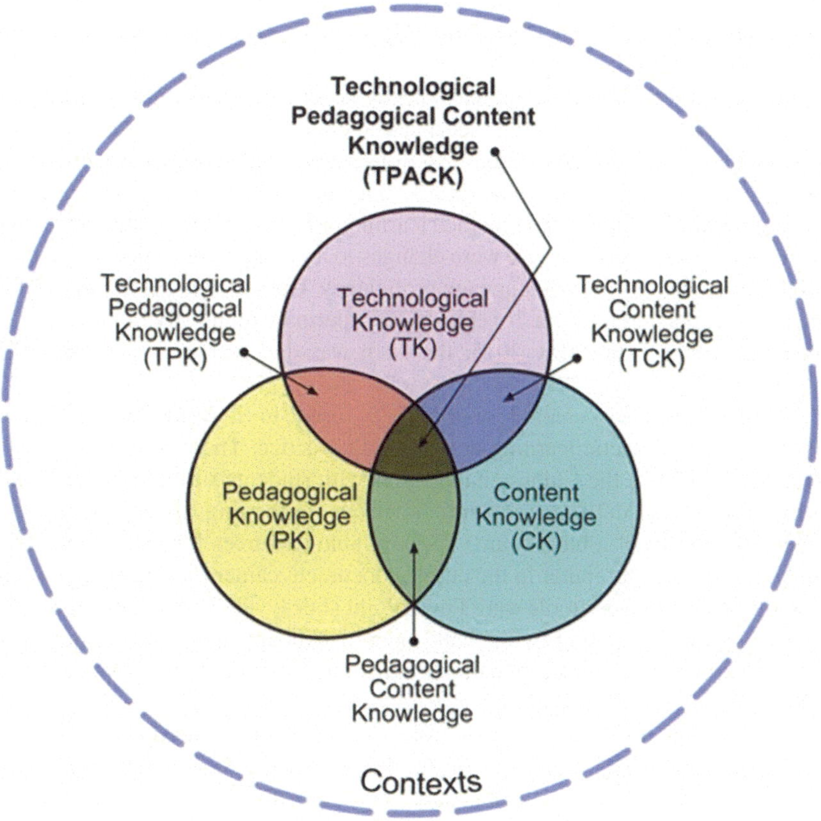

Fig. 7.4 TPACK framework as outlined by Mishra and Koehler (2008)

Questions in the other section are prefaced by the phrase: "How confident are you that you have the knowledge, skills and ability to support your students' use of ICT in the following…."

The survey of lecturers in the UTAS School of Engineering revealed their TPACK was significantly lower than that of preservice teachers at another Australian regional university.

Enhancing Student Learning in China

7.5 Teacher-Centered Learning

Starting in 2013, lecturers and postgraduate candidates from the UTAS School of Engineering delivered engineering units (subjects) in China at the Shandong University of Science and Technology (SDUST), the Xi'an University of Science and Technology (XUST), and the WESTA (West Australian and Tasmanian) college of the Southern University in Chongqing. All content was delivered face-to-face and online in the English language.

Until September 2016, all the teaching by Australian academics in China followed the traditional lecture-style, teacher-centered practice of face-to-face delivery. Each unit at SDUST comprised 72 h of technical content together with tutorial sessions and laboratory work in the ratio 39 h lectures/13 h tutorials/20 h laboratory (or design studio). Content and intended learning outcomes (ILOs) were identical to that delivered in Australia. There were changes to assessment in some units to compensate for students' English language proficiency. Up until May 2015, the 72 h in each unit was delivered over a 2-week (10-day) period with mostly 8-h face-to-face contact each day. From May 2015, the 72 h was delivered over a 3-week (15-day) period.

The term "teacher-centered learning" is used to describe the traditional Confucius-style, didactic learning and teaching practice. There were large classes (typically 70–100) at the Chinese universities with students sitting in rows of fixed seating. The teacher delivered content knowledge, in the English language, from a raised dais in front of a blackboard. The available resources for teachers included mainly a stationary computer in the class, a document camera, video projector, and a sound system. Delivery tools were PowerPoint slides, videos, and the emphasis of important points by writing (with chalk) on a blackboard. It was the norm that in-class engagement was limited to half a dozen students sitting in the front rows with the best English language skills.

The Australian teachers were responsible for the planning, designing, and delivering of content. In the first years of the delivery at the Shandong University of Science and Technology (SDUST), only very limited pedagogical guidance was provided from the more experienced teachers as they themselves had little or no experience of teaching in China. Hard copies of onshore lecture notes and tutorial problems were made available to the offshore teachers. These materials were delivered with little or no change, although the more experienced teachers were able to modify the teaching materials by, for example, including translations of technical terms.

There was a bilingual Chinese tutor in each of the classes. Their role was to record attendance and translate important details of timetables, class locations, assessment types, and dates. The tutors mostly had some knowledge of the subject being taught but adopted a passive role and did not contribute to the delivery.

Practical work was undertaken in adequately equipped laboratories with some apparatus and instrumentation copied from that used in Australia. Apart from laboratory instructions prepared by the Australian teachers and Australian-made equipment, all WH&S documentation in laboratories as well as equipment labeling was in the Chinese language.

7.6 Student-Centered Learning

One of the more experienced Australian teachers, Dr. Peter Doe, initially used a mix of "chalk and talk," PowerPoint slides, and videos for the delivery of content in China. English words that had a particular technical meaning were written on the blackboard. The meanings of technical words were explained, and sometimes a student was handed the chalk and invited to write the word on the board as a character and in pinyin. The teacher attempting to pronounce the word invariably brought forth howls of laughter.

In April 2016, an experienced pedagogue Dr. Seeta Jaikaran-Doe conducted a peer review of Dr. Doe's teaching in a class of over 100 students (Doe et al., 2018). She suggested the following learning strategies to encourage class participation and deeper understanding:

- Students should prepare questions and answers related to concepts that were previously taught. Then a student would name a friend who had to respond to one of the prepared questions.
- Students should work in groups. Each group was given a concept to discuss, and then when the time is up, the spokesperson for each group would explain the meaning and application of the concept.
- Students can be shown a word related to a concept. They write what they know about the word in relation to what they had previously learned. Students were then called upon to present what they had written.

Other techniques recommended by Dr. Jaikaran-Doe for effective student-centered learning were:

- Directing questions to individual students
- Handing a piece of chalk to a student and inviting them to write on the blackboard
- Inviting a student to come to the front of the class and explain important concepts (in English and/or Chinese)
- Using flashcards to reinforce technical words
- Referring questions to discussion groups

7.7 Effective Delivery of Engineering Units in China

A study was conducted in 2016 to determine what strategies are most effective for the Australian university academics to adopt for effective delivery of engineering units at Chinese universities. A mixed methods approach was applied to in-class feedback and student surveys. Three major themes, language, learning, and social, were identified. These themes were further explored in a quantitative survey of Chinese students newly arrived in Australia (Doe et al., 2018).

This study examined the experiences of Australian academics delivering engineering content at two universities (SDUST and XUST) in the People's Republic of

China. Two years into a 4-year engineering degree at their home universities, the Chinese students transfer to the University of Tasmania in Hobart to complete a Bachelor of Engineering with Honors degree in a further 2 years (2 + 2 program). These students move from a familiar and regimented life to a very different culture and learning style taught in a different language. Not surprisingly, many of the students perform poorly in their first semester of study in Australia.

The Australian academics teaching in China in the 2 + 2 program are typically confronted by 100 or more students sitting at fixed desks. Second year Chinese students' command of the English language is not sufficiently developed to be able to follow a typical Australian "lecture." The Australian academics are able to engage with the few students sitting in front of the class, but most of the students are not able to understand what is being said. Many try to follow the lecture notes that the lecturer had posted in advance on the students' learning website; the more assiduous students would have penciled in translations of technical terms. Some students use their mobile devices to translate technical terms while the lecturer is speaking. Toward the back of the classroom, students play computer games, text their friends, or fall asleep.

From the start of formal schooling, Chinese students are expected to be passive learners (Loh & Teo, 2017). Attendance in class is mandated. Students learn to respect their academics and not to question the validity of their teachings. The challenge for the Australian academics is to prepare these students for the very different student-centered teaching and learning style they encounter when transferring to Australia. Accordingly, the Australian academics attempt to engage the students.

A skilled lecturer can engage students by directing questions to the class or to individuals by name. However, this author found it difficult to engage with students in large classes in China. In 2015, in the Engineering Design and Project Management class at SDUST, he sought feedback from the class on a cartoon video on project management that had just been shown (Doe et al., 2018). He walked to the center of the class (105 students) and asked the question: "Who thought that video was interesting?" There was no response. "OK," he said, "who thought that video was funny?" Again, no response. After receiving no response to the third question: "Who has understood anything I have just said?" he realized that there was a major problem with engagement and communication.

He then directed the students to assemble into their laboratory groups (eight students in each) and after 15 min write three suggestions on the blackboard how to improve communication in this class. The results *verbatim* are shown in Table 7.1.

The results of an NVIVO autocoding function applied to students' responses to the in-class feedback (Table 7.1) are shown in Fig. 7.5. The coding words relating to *learning* identified in the responses were *study, questions, understand, classroom, example, teacher, and information.* The majority opinion of these students (still in China) was that the learning style of the Australian university academics was more important than the students' English language proficiency.

A review of literature identifies many factors affecting students' performance after transitioning to study in Australia including social and cultural differences,

Table 7.1 *Methods to improve communication*

Use some simple words and speak slowly. We study English
Use pictures instead of speaking English
Play some simple games to improve atmosphere
Have more passion and some body language
Emphasize the keyword
Ask students to express their idea after teacher tell us something
Dinner outside to discuss interests
Write down the complex words if necessary, and we can use the dictionary in the cell phone
The teacher can speak clearly, and I hope the teacher can give the students some time to discuss
Ask questions individually writing standard
We should boldly speak English; don't be shy
Write down what you say if you feel that we don't understand what you say
Can you talk with our Chinese teacher, if we don't respond to you, may our Chinese teacher to translate it
Make our class atmosphere happily and talk about with students frequently
Let us Chinese teacher translate the question if it is important
Speak more slowly and simple words please
Enrich classroom content, do many examples
Active classroom atmosphere, ask much question
Keywords write on blackboard
Use some body language if we don't understand you
Use some simple words or pictures to replace difficult words
More communicate with students
Give us many vivid example
We could understand by using easily words
Using body language makes us understand what you say
We think teacher should participate more in the production process and give more advice
It would be easier for us if you provide some webs for us to search information

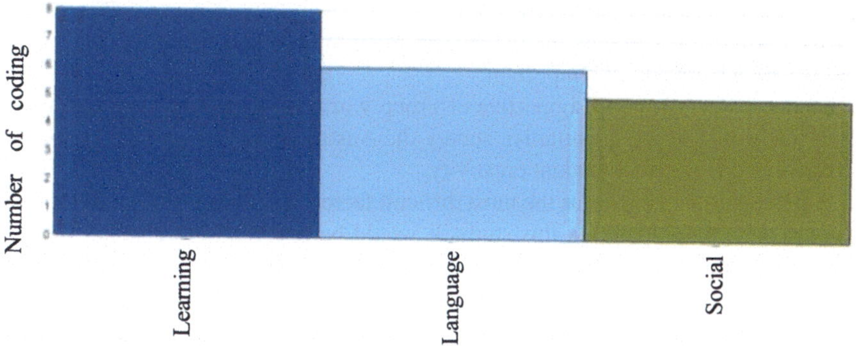

Fig. 7.5 The number of references related to learning, language, and social

lack of technical language fluency, and the change in learning and teaching (L&T) styles.

Lyden (2016) surveyed Chinese students (N = 22) who had transferred to Australia and were enrolled in the third and fourth year units. These students had previously completed 2 years of an electrical engineering degree course at SDUST. Although this study identified some of the reasons for students' poor performance in their first semester in Australia, it also probed their learning and teaching experience at Chinese Universities before coming to Australia. One question relevant to the subject of this chapter was:

- *What expectations did you have before coming to Australia to study? What has been most challenging since you arrived?*

The Chinese students' responses to this survey question are shown in Table 7.2. The same NVIVO autocoding (as for the in-class feedback shown in Fig. 7.5) was used: *language*, *learning*, and *social*. The results are shown in Fig. 7.6.

It should be noted that the students participating in this survey had already transferred to the Australian university and had experienced life in Australia and study in the Australian university engineering course. Although the first part of the survey question "What expectations did you have before coming to Australia to study?" related to the students' experience at SDUST, the second part of the question "What has been most challenging since you arrived?" was not relevant to the students' experience in China. However, it is revealing that these students reported that language was more important than learning styles.

The principal factors that limit the effective delivery of Australian university engineering units at the Chinese partner universities were found to be communication (*language*) and teaching style (*learning*). Before they transferred to Australia, the teaching style was of more concern. After transferring to Australia, however, the Chinese students discovered that their English language proficiency was not adequate, in particular their understanding of technical English. This conclusion (from the qualitative research) was not supported by a quantitative survey administered to students a few months after articulating to the Australian university (Doe et al., 2018). Perhaps the students' experience in Australia was such that both *language* and *learning* were equally important.

Students were strongly supportive of group work. Also, there is some evidence from the results of the quantitative survey the Australian university lecturers were effective in encouraging student creativity.

Adjusting to language was the most difficult factor for the newly arrived Chinese students who participated in this study. It would be interesting to conduct further research to identify exactly what aspects of language the students find the most challenging and what mitigation strategies are most effective. Would bilingual academics be better communicators? Is the problem a lack of fluency in technical English, the speed at which the lecturers and fellow classmates talk, or difficulty in adjusting to different accents?

This research has confirmed the widely held view that lack of fluency in the English language is the prime cause of poor performance of Chinese students in

Table 7.2 *Responses to questions: What expectations did you have before coming to Australia to study? What has been most challenging since you arrived?*

Difficulties fall into getting across jargons which are involved in the class and the jokes which (Australian teacher) has made
Honestly, I expected a friendly study environment in (Australian university), and I did have a quite good study experience so far. I think most local students are very nice and friendly and ready to help us. I think language and culture difference are the most challenging things
I wanted to study a new culture, and the difficult is language
I expected for working with local students all the time; the results are good enough. I consider the most challenging thing I ever meet is getting along with other roommates in the rented house
Lack of fundamental knowledge of first 2 years' units and English language!!!!!! I think the most challenging is English!!!! Some tutor said
Language, I think I can understand what the people say, but it is not. And I cannot understand the professional engineering words when having lectures
The expectation I have before coming here is to get a better learning process and getting more advanced knowledge. Yes, (this university) provided me sufficient of those I mention. The most challenging thing I met is the education methods, the content is not the same so there is a lot of prerequisite that I need to learn by myself which make my first semester terrible because I have to keep up with the lectures up to date and also to learn the knowledge previously the classmate already learned
The most difficult thing is language barrier and I expect to learn more useful knowledge which is not learned in China
To be honest, the hardest challenge i have met is that whatever methods I tried, there are lots of contents that I can't understand and I can't explain it by myself directly. But this situation become much better and i need to thanks teacher's helps. Before coming to AU, I have a strong ambition about improve English in a short period and can touch more up-to-date knowledgeable, but everything need time to make effect on
1. Get good grades 2. Language and food. Poor language
English problem
1. I want learn more knowledge in Australia. 2.The biggest challenge in language problem, I can't understand all the words in class because there are many professional words
Environment has changed
Improve me in professional knowledge, language success in study, in life arranging and communication

their first semester studying engineering in Australia. Engaging with the students has proven difficult due to large class sizes in China and the ingrained culture of passive learning. Australian academics teaching in China should use the strategies put forward by Dr. Jaikaran-Doe described above to create a student-centered environment in class to better prepare them for study in Australia. One very effective technique is to encourage group work—students learn from their peers. Besides, there is not the fear of failure to communicate when one student is supported by others in his or her group.

Chinese students in their first semester in Australia rate language proficiency as more important to successful study than lecturers' teaching style. This observation led to the appointment of an experienced pedagogue (Dr. Seeta Jaikaran-Doe) to work in China teaching English language focusing on writing engineering

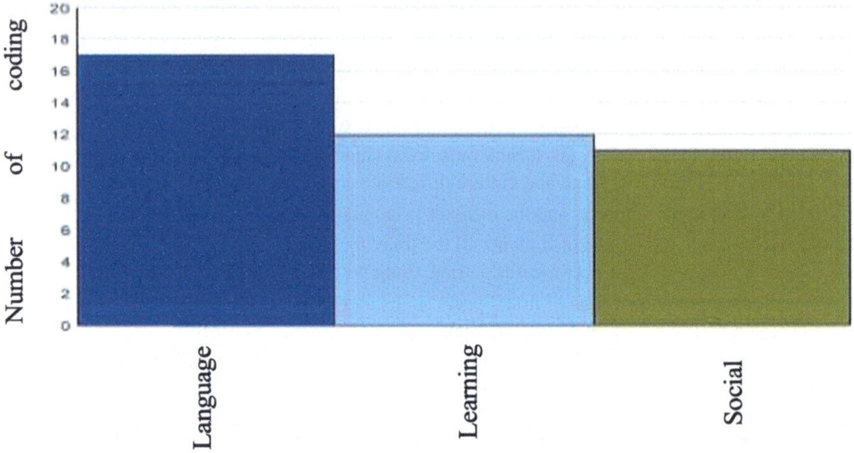

Fig. 7.6 The number of references related to the three identified major themes: learning, language, and social

laboratory reports, gaining better knowledge of technical English, and developing oral communication skills.

This study was the start of research focusing on a better outcome for Chinese students entering the 2 + 2 program at the University of Tasmania. The methods developed for Australian academics to engage students with a spectrum of English language skills in large classes, together with a program of support for newly articulated students, have proven to be effective (Jaikaran-Doe et al., 2018).

7.8 Articulation of Chinese Students into the UTAS Bachelor of Engineering Program

The majority of newly arrived Chinese students performed poorly in their first semester in Australia. This is despite some students living in Australia and participating in English language study at the university's English Language Centre for up to 6 months before admission to BE (Hons) degree courses.

In 2015, a total of 16 electrical engineering students, who had completed 2 years at SDUST which included units delivered by the UTAS lecturers, transferred to Tasmania and enrolled in the UTAS Bachelor of Engineering course in Hobart, Tasmania. These students were first enrolled in an English language program at UTAS for 3–5 months, depending on their International English Language Testing Scheme (IELTS) level, from August 2014. At the end of their first semester in the engineering degree course at UTAS, only three of these students had passed all units. Ten students failed one or more units, and three students failed all units in their first semester.

In July 2016 another cohort of 19 civil engineering students from XUST were enrolled by direct entry, having already satisfied the level of English proficiency required for their student visa. A framework for orientation and support of these students was prepared by Dr. Seeta Jaikaran-Doe who had been delivering an English language course at XUST. At the end of their first semester in the engineering degree course at UTAS, nine of these students had passed all units. Eight students failed one or more units, and three students failed all units in their first semester.

The direct entry students arrived in Australia only a few days before the start of the semester. Prior to their arrival, the UTAS engineering staff attended a meeting at which two staff recounted their recent experience delivering second year civil engineering units to these students at XUST. This meeting helped alert other lecturing and tutorial staff to a situation where for the first time a substantial proportion of their third year civil engineering classes (circa 40%) would be newly arrived Chinese students.

The poor performance of the 2015 and 2016 cohorts prompted the provision of resources for a support framework for communication (Fig. 7.7). Six 1-hour sessions to plan, design, and implement strategies to support the students were delivered to newly arrived civil engineering students from XUST by Dr. Seeta Jaikaran-Doe. Simultaneously, tutors who were fluent in Mandarin and English provided a 1-h tutorial per week for each unit in which the students were enrolled. Dr. Jaikaran-Doe conducted a special session on pedagogical strategies to guide these tutors in the delivery of their tutorials.

Activities for the Chinese students were not conducted in isolation. Since they would be working and studying with domestic students, it was important to involve the domestic students in some of the activities. Dr. Jaikaran-Doe conducted three separate in-class sessions varying from 45 to 30 min to students in the third year civil engineering units. The activities were designed to reduce the cultural divide between the two cohorts of students, encourage communication, and bring an

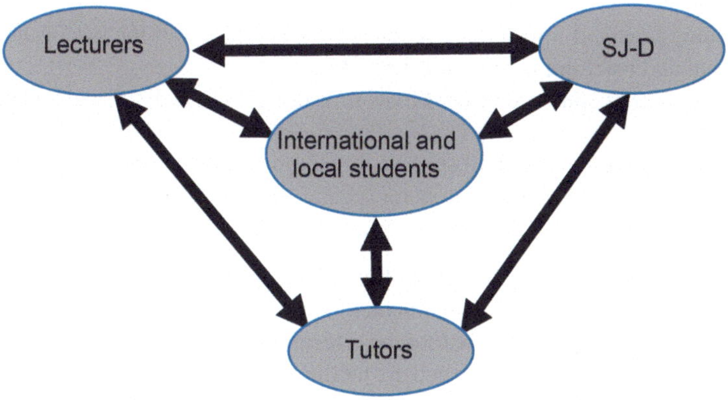

Fig. 7.7 Support framework for communication (Jaikaran-Doe et al., 2018)

atmosphere of collegiality. The intended learning outcomes (ILOs) for these activities were:

- Bridging the gap between domestic and international students
- Informing international students about expectations for teaching and learning at the Australian university
- Enhancing communication among tutors, lecturers, international students, and local students
- Providing practical information to international students of the use of the online learning and teaching facility at the Australian university

In addition, four separate 1-h sessions were conducted for only the Chinese students by Dr. Jaikaran-Doe to inform them about plagiarism, academic misconduct, work health and safety, assessment procedures, and how to access and use the learning management system, MyLO. An "International Student Induction Manual" detailing important matters that students need to know about the academic culture of the school was distributed and explained to them. WeChat was used for easy communication among tutors, students, and Dr. Jaikaran-Doe.

The intervention strategies were extended in 2017 to include training workshops for both international and domestic tutors. Dr. Jaikaran-Doe was appointed to a part-time position within the UTAS School of Engineering as International Transitions Officer with the responsibility of recruiting, rostering, and providing pedagogical advice to the Mandarin-speaking tutors.

The significantly improved performance of the Chinese students in their first semester at UTAS supports the proposition that the strategies in the support framework addressed the pedagogical factors leading to the high failure rate of the previous cohort; these were:

- Bridging the gap between domestic and international students
- Informing international students about expectations for teaching and learning at the Australian university
- Enhancing communication among tutors, lecturers, and international and local students
- Providing practical information to international students of the use of the online learning and teaching facility at UTAS (MyLO)
- The adoption of some Intensive Mode Teaching (IMT) techniques by some staff and tutors
- Reviewing the progress of all students in all units midsemester
- Engaging Chinese-speaking tutors for special additional tutorial support throughout the semester

The implementation of the support framework also has resulted in engineering lecturers at the University of Tasmania making their lecture notes and other materials available to students online before the relevant class.

A buddy system was established to facilitate interaction and communication between international and Australian students particularly during group work. Local and international students were invited to write a report of approximately 500 words

SDUST S2 2016

78	72	81	73
61	61	57	72
61	54	60	55
56	57	60	52
56	55	61	39
50	39	34	38
56	35	32	26
39	49	46	48
AN	26	30	36
37	42	36	36

INTERVENTION

SDUST S2 2017

67	80	75	50
66	67	50	50
54	50	71	58
66	76	68	60
66	73	90	50
68	72	68	50
71	66	58	N*
60	70	76	48
44	73	64	42
44	76	64	33
47	65	N*	43
39	49	N*	40

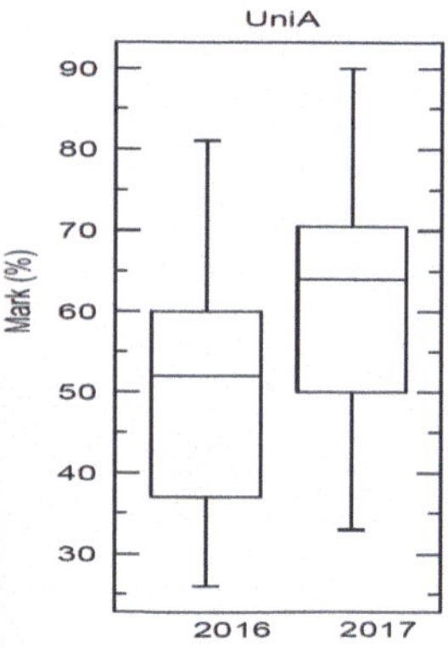

Fig. 7.8 End-of-semester examination results for all cohorts from SDUST. *N* failed supplementary examination, AN* absent from examination. The probability of this result, assuming the null hypothesis, is 0.0010*

of how he/she has supported or contributed to an international student's well-being. An award was presented for the best buddy outcome.

Chinese students developed more confidence to communicate with teachers and tutors to discuss matters affecting them.

The overall performance of electrical engineering Chinese students from SDUST in 2017 in their first semester of study at the University of Tasmania was significantly better than that of a similar-sized cohort students from SDUST arriving in 2016 despite the latter having transferred up to 6 months earlier to study English Language (Fig. 7.8). The better performance of the 2017 students is attributed to the support framework.

The effectiveness of the support framework for the civil engineering students from XUST is shown in Fig. 7.9 with second semester results being significantly better than first semester results.

7.9 Intensive Mode Teaching in China

The Intensive Mode Teaching (IMT) technique resulted from a major Australian Office of Learning and Teaching program involving a number of Australian universities focusing on engineering education (Male et al., 2016). The Intensive Mode Teaching (IMT) subject matter is typically delivered in blocks of 4 h or more

XUST S1 2017

64	50	76	52
66	55	52	62
60	61	57	52
58	54	52	65
74	57	60	51
71	74	66	74
65	58	55	W*
57	23	54	58
65	58	55	W*
62	50	62	48
68	40	68	W*
56	50	43	W*
55	39	41	W*
40	50	42	W*
60	34	55	45
46	29	39	W*
60	39	48	44
54	42	43	40
41	41	44	39
W*	W*	W*	W*

XUST S2 2017

66	76	61	68
56	53	53	62
59	69	60	65
61	60	50	56
68	77	73	82
63	57	50	54
62	72	50	61
69	71	61	62
64	56	52	55
72	84	65	76
53	71	53	55
66	73	68	61
56	55	35	52
56	50	31	54
50	62	45	51
44	52	53	60
46	77	76	58
41	53	36	40
46	49	33	47

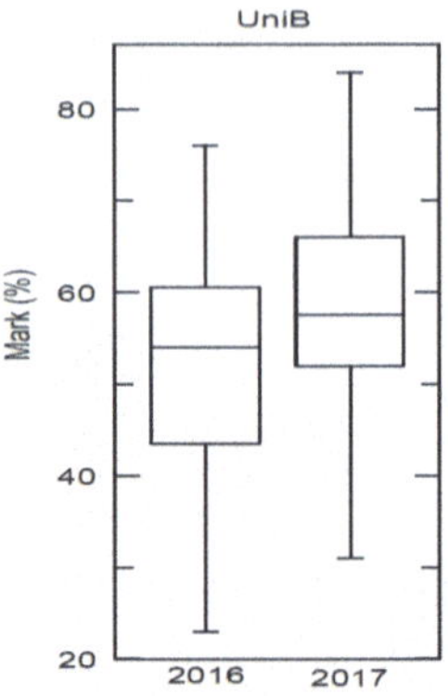

Fig. 7.9 End-of-semester examination results for cohorts from XUST. *The probability of this result, assuming the null hypothesis, is 0.0058*

delivered over a period of weeks, as distinct from the usual delivery spanning a term of several months. There is emphasis on "threshold concepts." These are the fundamental principles on which a body of knowledge can be built. IMT relies on students having access to, and studying, subject material prior to face-to-face interaction. In that way, the teacher can concentrate on imparting "threshold concepts."

Interactive Intensive Mode Teaching (IMT) techniques were used to deliver an engineering design and project management unit over a period of 6 weeks to large classes (75 and 115 students) in May–June 2017 at the Shandong University of Science and Technology (SDUST) in China.

A study was conducted to examine the effectiveness of the interactive IMT technique, and the assessment tasks used for the delivery of a project management unit by comparison with the traditional, didactic style of teaching employed by the same teacher (Dr. Peter Doe) in 2016.

The timetable for each cohort was 4 h in a lecture room and 8 h in a design studio each week. The teaching load for the UTAS teacher was 24-h contact each week. Chinese university lecturers were present in the classroom and design studio most of the time. All delivery was in the English language except for announcements (in the Chinese language) by the Chinese lecturers.

Doe et al. (2017) surveyed newly arrived civil engineering students in their first semester in the third year of a civil engineering degree at the University of Tasmania. The survey probed the students' experience of learning and teaching experience in China, in particular in the classes delivered in China by Australian academics. That study identified the different learning and teaching styles of the Australian teachers as being of most concern to students while they are still in China, but after coming to Australia it is their language that is of most concern—particularly understanding technical terms. The delivery of engineering content using interactive Intensive Mode Teaching was thought to have the potential to assist these students with their transition to UTAS.

In 2016, prior to the introduction of an online learning management system (Cloudcampus) at the Chinese university (SDUST), a unit *Engineering Design and Project Management* was delivered by Dr. Peter Doe in the traditional teacher-centered style with the teacher delivering content knowledge from a raised dais in front of a blackboard to approximately 100 students sitting in rows of fixed seating. The ICT resources available were a stationary computer, a document camera, video projector, and a sound system. Slides, videos, and the emphasis of important points by writing (with chalk) on a blackboard were the major delivery tools. Engagement with students was limited to those students who chose to sit in the front rows and exhibited the best English.

The following IMT techniques were employed in the delivery of *Engineering Design and Project Management* at the same Chinese university a year later in May/June 2017:

- All lecture material was uploaded onto Cloudcampus before the start of teaching.
- In-class sessions were limited to 4 h/week to maximize time in the design studio.
- Content in the uploaded notes was *not* repeated in class (e.g., the explanation and function of the Raspberry Pi General-Purpose Input/Output port, also programming in Python and the EV3 LEGO robot languages).
- An online formative assessment (Quizizz.com) preceded each fortnightly summative test, as was found valuable in a unit on critical theories of technological development and an intensive accounting unit delivered elsewhere.
- Student engagement was encouraged both in-class and in the design studio through group competition.
- Time in-class was allowed for groups to discuss threshold concepts and report verbally or by writing on the blackboard.

7.9.1 Intended Learning Outcomes

The following statement on ILOs was approved by the UTAS School of Engineering and ICT's Learning and Teaching Committee and included in the *Engineering Design and Project Management* Unit Outline:

On completion of this unit, students should be able to:

- Demonstrate an understanding of the fundamentals of engineering design, project management, engineering ethics, and sustainability
- Develop and evaluate conceptual designs to address the problem specifications
- Function as an effective team member using simple project management tools to build and test a prototype design
- Critically reflect on your and others' contributions to the team and the execution of the project

7.9.2 Assessment Tasks

Assessment tasks for *Engineering Design and Project Management* delivered in 2017 comprised a group video report on the design-and-build component together with fortnightly in-class tests and a one-page, handwritten diagram (with explanation) of one of the microprocessor-based circuits built and tested in the design studio.

Table 7.3 details the assessment tasks and their alignment to the unit's ILOs and Engineers Australia (EA) Elements of Competency for Stage 1 Professional Engineers (Engineers Australia, n.d.).

Graduates satisfying the stage 1 competencies for professional engineers is part of the requirement for accreditation of engineering schools at Australian universities. The competencies the unit *Engineering Design and Project Management* claims to deliver are listed below:

1.3 In-depth understanding of specialist bodies of knowledge within the engineering discipline.
1.4 Discernment of knowledge development and research directions within the engineering discipline.

Table 7.3 Alignment and weighting of assessment tasks, ILOs, and EA competencies

Component	Weight	Due date	Addresses ILOs	Addresses EA competencies[a]
1. In-class tests (individual)	36%	Fridays weeks 2, 4, and 6 in class	1	1.3 and 3.1
2. Circuit diagram (individual)	14%	End of week 6	2	1.4, 2.1, and 3.3
3. Project (team)	50%	End of week 6	3 and 4	2.4, 3.5, and 3.6

[a]Relevant Engineers Australian Stage 1 Competencies

2.1 *Application* of established engineering methods to complex engineering problem-solving. Vital aspects of an engineer's skill encompass design, innovation, and project management. It is essential that an engineer can work within a framework of professional standards and yet have the innovative skills to produce creative design solutions. Students submit two teamwork assignments.

2.4 *Application* of systematic approaches to the conduct and management of engineering projects. Students develop an understanding of the fundamental concepts of engineering design, through studying the design of everyday artifacts as well as professional design practices. Students are introduced to contemporary concepts of innovation and project management within a competitive environment.

3.1 *Ethical* conduct and professional accountability. Students are introduced to the Engineers Australia Code of Ethics.

3.3 *Creative, innovative, and proactive demeanor.* Students apply creative approaches in a design situation.

- *3.5 Orderly management of self and professional conduct.* Students have multiple tasks to prepare and report on a strict deadline with penalties for late submission. This necessitates prioritization given the requirements of other units (Engineers Australia, n.d.).

7.9.3 Online Learning Management System

- SDUST introduced an online learning management system (Cloudcampus) in time for the delivery of the *Engineering Design and Project Management* unit in May/June 2017. The Unit Outline and all lecture notes were uploaded 2 weeks before the start of teaching. Over each weekend, the teacher uploaded the content to be delivered the following week with sections highlighted to show the threshold concepts which were to be discussed during the in-class sessions. In-class, the lecturer projected the content as a MS Word document (rather than PDF or PowerPoint) so other threshold concepts could be highlighted at the time if needed. Even though both the engineering content and the English language teachers stressed the importance of regularly reading postings on Cloudcampus, only 45% of the students enrolled in *Engineering Design and Project Management* accessed the information in Cloudcampus. However, it should be noted that Chinese students enjoy strong peer support through living together on campus. Content downloaded from Cloudcampus could have been shared.

7.9.4 Results

Table 7.4 shows a comparison of results for the summative assessments in *Design and Project Management* ($N = 186$) delivered in 2017 (using the IMT techniques) with the results of a similar cohorts ($N = 194$) who took the unit *Engineering Design*

Table 7.4 Components of students' overall performance

Assessment task	2016 Mean (%)	2016 Standard deviation (%)	2017 Mean (%)	2017 Standard deviation (%)
In-class tests	85.9	12.8	77.8	12.8
Design project group video report	70.4	10.4	5.7	8.5
Circuit diagram	n.a.	n.a.	50.0	29
Overall result	80.2	9.2	71.7	9.6

and Project Management in 2016, delivered by the same teacher using teacher-centered, didactic teaching methods.

An independent samples *t*-test was conducted to analyze the students' results. *Cohen's d* was computed to identify and verify the magnitude of the differences between the means of the 2016 and 2017 results.

The overall result for the 2016 students ($M = 80.2\%$, $SD = 9.21\%$) was significantly better ($p < 0.0001$, $t = -4.37$) than the results achieved in 2017 using the interactive IMT techniques. *Cohen's d* was 0.90 indicating a "large effect size." This result was unexpected. The reasons for this unexpected result are explored below.

- Feedback from one student on the trial of IMT techniques in a final year engineering unit at UTAS in March 2017 suggested that the expectation of students was for a lecturer to "teach" them (P. Doe personal communication June 30, 2017). The focus of student-centered learning is for the students to "learn" rather than be taught. If a final year engineering student in Australia wants to be "taught," it could be that the acceptance of IMT in China would be problematical considering Chinese students' history of didactic learning. It may also be that interactive Intensive Mode Teaching techniques have only recently been introduced in the delivery of engineering units at UTAS and elsewhere. Students develop an expectation based on their prior years of study, and when something different is introduced, they are resistant.

7.9.5 Technical English Fluency

- Teaching and learning in the design class was constrained by students' lack of technical English fluency. Many students did not know the English word for "screwdriver." A student in the design studio looked blankly when the teacher offered to "fix" a problem. Other students chose to download Chinese language versions of software making instruction by the UTAS teacher difficult. It is possible that students' inability to read and understand the content uploaded in English to the online learning management facility is also a limiting factor in the effectiveness of interactive IMT in China.

7.9.6 Synergy with English Language Program and IMT Trial Outcome

During the 2017 delivery of the content *engineering design and project management*, there were instances where synergies occurred in the form of students' use of vocabulary from an *Academic Skills Program* concurrently being delivered at SDUST by Dr. Jaikaran-Doe. As an example, in the delivery of content on *engineering ethics*, the teacher identified "bias" as a threshold concept. After allowing a few minutes for the design class groups (eight students in each group) to discuss the concept, the teacher handed the chalk to one of the students who wrote an example of "bias" on the blackboard. The student wrote: "An example of bias could occur at a job interview when the applicant was not appointed on the basis of race." The teacher was surprised by the vocabulary until it was pointed out that students had recently participated in mock interviews in their *Academic Skills Program* class.

Notwithstanding the authors' enthusiasm and initiative in embracing student-centered learning and interactive IMT techniques, the effectiveness of delivery of the *Engineering Design and Project Management* at the SDUST in 2017 (as measured by students' results) was disappointing. This experience further confirms the authors' observations that students' technical English language skills present a challenge to effective delivery of engineering content. However, some IMT techniques proved to be effective as evidenced by students' in-class participation in the discussion of threshold concepts.

The University of Tasmania is revising the curriculum for its Bachelor of Engineering with Honors degree. The revised curriculum includes a significant component of experiential, project-based learning and design project across all specializations. If student-centered learning and teaching is to be embraced in the new curriculum, it is vital to explore further the reasons for the disappointing results from this introduction of interactive Intensive Mode Teaching in China.

7.10 Strategies for Promoting Cultural Diversity

It is mandatory for engineering graduates (Stage 1 Professional Engineers) to demonstrate effective team membership and team leadership (Engineers Australia, n.d.). Therefore, Australasian engineering degrees are typically organized to incorporate this competency as part of course learning outcomes. Instructional activities are developed for students to achieve this competency, for example, laboratory and design studio sessions with group reports and presentations are typically used to assess the level of competency attained.

It is one of the requirements of a graduate engineer that he or she is an effective team member and "Functions as an effective member or leader of diverse engineering teams, including those with multi-level, multi-disciplinary and multi-cultural dimensions" (Engineers Australia, n.d., p. 6). In the context of Australasian engineering degrees, cultural diversity manifests primarily via a student body which consists of both a domestic cohort and an international cohort. Thus, there is usually ample opportunity for cultural diversity within teams, particularly during laboratory activities. Unfortunately, however, students tend to form groups with little cultural diversity if allowed to self-select. That is, groups mostly consist only of domestic students or only of international students. Indeed, domestic students have on occasions been observed to actively resist lecturers' attempts to mandate cultural diversity within groups (Henderson, A., personal communication, December 10, 2017).

What usually happens, when self-selection for group formation is followed, is that students form groups with their friends. Generally, this method facilitates harmony between group members, but the various advantages afforded by cultural diversity are then largely foregone. The most common alternative is some means of random selection, which increases the likelihood of cultural diversity within groups but which may potentially do so at the expense of harmonious teamwork.

It seems to be mostly recognized by students and lecturers that cultural diversity is a positive thing and also that current "default" methods of group selection generally prevent that. UTAS lecturers are actively pursuing/trialing methods to achieve diversity but with mixed results and clearly with some disharmony being apparent. A probable failing of many groups is that they simply divide the work up piecemeal and then join it together at the end, often resulting in a poor final product, rather than realizing that one person in the group is better at writing, one better at organization and managing tasks, one better at data analysis, and one better at maths. There is a need to provide students with training on how to work in a group.

A survey of local students (N = 4), international students (N = 4), and staff (N = 3) reported by Jaikaran-Doe et al. (2018) concluded diversity in groups has value, but it is not generally yet being taken full advantage of. Group formation should only be considered the first part of the process, and to ensure that culturally diverse groups lead to positive outcomes and experience, it is likely necessary for teaching staff to provide support and suggested structure for those groups, for example, to help the members recognize the differences and individual strengths within a group and thus capitalize on them.

Discussion at the 2018 Australasian Conference on Engineering Education (Jaikaran-Doe et al., n.d.) in Hamilton, New Zealand, canvassed the views of the importance of cultural diversity and group formation. The survey and workshop respondents were quite clear on the meaning of cultural diversity. A major proportion of the respondents (82%) were of the opinion engineering laboratory and design groups benefitted from cultural diversity. A majority of respondents (63%) favored some form of group formation other than self-selection.

Allowing students to select their partners and then putting the partners together to deliberately form cross-cultural groups worked quite well in the opinion of one lecturer. Conversely, one domestic student advocated self-selection so that the more academically able students could work with their peers to get good results. Groups formed on the basis of a common interest in specific tasks reportedly worked well. Students tended to be task-focused rather than being able to perceive the benefits of working in a culturally diverse team.

A number of other factors touching on harmony within culturally diverse groups emerged from the NVIVO coding of the discussion. Group size, team roles, language, communication, and assessment were examined. Group size was very much determined by class size. Larger institutions adopted group sizes of 10 or more, while the smaller regional institutions favor groups of four to five students. Assessment, particularly the assessment of individual students within groups, emerged as a contentious issue. Not unexpectedly, language and communication emerged as the most significant issue impacting on harmony within culturally diverse groups. Respondents were uniformly of the view that language and communication play a vital role in the harmony and effectiveness of group work.

The conclusion from analysis of the AAEE 2018 Conference discussion was that diversity in groups has value, but it is not generally yet being taken full advantage of. Group formation should only be considered the first part of the process. Assuring cultural diversity in groups can lead to positive outcomes, but teaching staff must provide support and suggest structure for the groups and help the members recognize the differences and individual strengths within a group. Of vital importance is the provision of English language support for students who have difficulty in communicating.

7.11 Continuous Improvement

New teaching principles were approved at the University of Tasmania in June 2023. The principles are based on providing a structured approach to self-assessment, allowing educators to reflect on and improve their teaching practices, providing a clear and consistent basis for collegial peer review for teaching practice, and offering a comprehensive foundation for expert review. The guidelines do not provide examples of, or prescribe, modes or styles of delivery; they are more about the management of teaching. For example, in the domain "Student-Centered Teaching Delivery" with the sub-domain "Teaching Delivery," there is a range of entries ranging from "As a developing teacher you: Have an understanding of and use spaces or technologies appropriately to facilitate student learning" to "As a leader in learning and teaching you: Lead innovation in enhancing quality teaching practices and supporting student learning at university, disciplinary, and (inter) national levels." Without explicitly stating, the levels appear to be aligned to the levels of seniority within the university from associate lecturer to professor, although excellence in teaching is only one criterion for promotion. Besides, associate lecturers may be (and often are) excellent teachers.

These teaching principles have been adopted university-wide. They apply across the spectrum of disciplines from fine arts to medicine. As detailed above, engineering has particular needs with respect to laboratory-based learning and the development of the so-called soft skills, communication, leadership, and management.

Wrobel-Tobiszewska et al. (2023) prepared a set of guidelines for blended delivery of engineering units (Table 7.5). These guidelines apply to modern delivery styles and methods online, face-to-face, and combinations of these.

Table 7.5 Elements of an ideal blended learning environment

Welcome students to the unit	Introduce teaching staff
	Establish communication channels details
	Provide assessment task details
	Encourage the use of web cameras during web conferencing
Establishing and maintaining social presence in an online environment	Actively contribute to online discussion
	Prompt online responses
	Building rapport by adding a human component to online interactions
Online student-lecturer communication	Class announcements
	Online discussion forums (ODF) Online chat apps
Prerecorded video lectures	Content chunking
	Interactive videos
	Existing video resources
	Other video types
	General video production tips
Other asynchronous learning activities	Asynchronous consultation
	Embedded activities
Live-online sessions	Lecturer's performance
	Student participation
	Relevance
	Technology
Enhancing face-to-face sessions (practicals, tutorials, workshops)	Hybrid synchronous sessions
	Sharing work on a whiteboard
Assessment tasks	Online workshop-based assessment
	Online discussion and reflection-based assessment
	Dependency links between assessments and learning material
	Video assessments
	Online group presentation
	Peer-reviewed assessments
	Online peer assessment
	Weekly contribution to an assessment
Feedback	Written feedback
	Audio or video feedback
	Student-teacher feedback
Techniques for improving student-teacher engagement	Quick to introduce
	Moderate effort required
	Additional effort required
Strategies for encouraging colleagues to engage	Barriers
	Motivators
	Positive actions to facilitate engagement of colleagues

7.12 Chapter Endnote

This chapter has detailed one example, the UTAS delivery of engineering units in their 2 + 2 program in China, of how engineering education has evolved over the past decade from the traditional teacher-centered style to modern student-centered methods embracing ICT and blended (online and face-to-face) learning. Applying Intensive Mode Teaching techniques in China was not as successful as had been hoped, but strategies to support Chinese students when they transferred to Australia had a very positive impact on students' results. More generally, in the opinion of participants in Australasian Association of Engineering Education conferences, having students work in groups was found to benefit learning particularly if cultural diversity in groups can be facilitated.

References

ClassroomAdvancements.(2017).4TypesofLearnersinEducation.AdvancementCourses.[Accessed 21.09.23]. https://blog.advancementcourses.com/articles/4-types-of-learners-in-education

Doe, P., Jaikaran-Doe, S., Lyden. S., Liu, M., Ren, B., Yang, P., & Male, S. (2017). *Intensive Mode Teaching for the delivery of engineering content to students at a Chinese University.* AAEE2017 Conference, Manly, Sydney, Australia.

Doe, P. E., Lyden, S., Jaikaran-Doe, S., & Wang, X. (2018). Enhancing students learning in an Australian 2+2 undergraduate engineering program. *International Journal of Higher Education, 7*(5), 86–95.

Engineers Australia. (n.d.). *Stage 1 Competency Standard for Professional Engineer.* Retrieved 24.10.23 from https://www.engineersaustralia.org.au/sites/default/files/2022-07/stage-1-competency-standard-professional-engineer.pdf

Finger, G., Jamieson-Proctor, R., Cavanagh, R., Albion, P., Grimbeek, P., Bond, T., & Lloyd, M. (2013). Teaching teachers for the future (TTF) project TPACK survey: Summary of the key findings. *Australian Educational Computing, 27*, 13–25.

Jaikaran-Doe, S., & Doe, P. (2015). Assessing technological pedagogical content knowledge of engineering academics in an Australian regional university. *Australasian Journal of Engineering Education.* https://doi.org/10.1080/22054952.2015.1133515

Jaikaran-Doe, S., Lyden, S., & Doe, P. (2018). Articulation of Chinese students into an Australian engineering degree. In *Proceedings of the 29th Australasian Association for Engineering Education Annual Conference (AAEE 2018), 09–12 December 2018, Hamilton, New Zealand.*

Jaikaran-Doe, S., Doe, P., Henderson, A., Lyden, S., Khan, Y., Kutay, C., & Symes, M. (n.d.). *Strategies for promoting cultural diversity within student laboratory groups in engineering degree courses in Australasia (unpublished manuscript).*

Loh, C. Y. R., & Teo, T. C. (2017). Understanding Asian students learning styles, cultural influence and learning strategies. *Journal of Education & Social Policy, 7*(1), 194–210.

Lyden, S. (2016). *International student perceptions and experiences of studying their final two years of engineering in Australia*. Unpublished manuscript.

Male, S. A., Baillie, C., Hancock, P., Leggoe, J., MacNish, C., Crispin, S., & Alam, F. (2016). *Intensive Mode Teaching Guide*. Retrieved from: Intensive mode teaching website: http://www.uwa.edu.au/imt/guide/

Mishra, P., & Koehler, M. J. (2006). Technical pedagogical content knowledge: A framework for teacher knowledge. *Teachers College Record, 108*(6), 1017–1054. https://doi.org/10.1111/j.1467-9620.2006.00684.x

Mishra, P., & Koehler, J. M. (2008). Introducing technological pedagogical content knowledge. In *Annual Meeting of the American Educational Research Association,* New York.

Shulman, L. (1986). Those who understand: Knowledge growth in teaching. *Educational Researcher, 15*(2), 4–14.

Turner, L., Wood, D., Jaikaran-Doe, S., & Doe, P. (2015). Investigating students' engagement in an engineering degree through technology enhanced learning. In *IFIP Conference 2015, Vilnius, Lithuania*. July 1–3.

Wrobel-Tobiszewska, A., Lyden, A., Holloway, D., Henderson, A., Doe, P., & Millar, B. (2023). Chapter 2 engagement in the blending learning modal of a post-COVID university. In H. Epinosa & I. Gratchev (Eds.), *Advancing engineering education beyond COVID: A guide for educators*. CRC Press.

Chapter 8
Construction of LEM-LT Framework

Abstract In keeping with my motivation for authoring this book, that is, to publish a part of Dr. Seeta Jaikaran-Doe's PhD thesis, I have included in this book, as this chapter, her framework for quality teaching and learning in the twenty-first century. It was developed from extensive research into the TPACK of high school teachers, but as reported in Chap. 7, it is directly applicable to learning and teaching at university level. The framework is based on the "systems level" attributes: support, resources, infrastructure, time allocation, curriculum redesign, and professional development together with "design principles" which include informative assessment, creativity, innovation, access to technology, and social and emotive contact with students. The framework clearly shows the alignment of the Levels of Teaching Innovation and TPACK and how in combination the students will acquire the skills for learning and living in the twenty-first century.

8.1 Introduction

This chapter presents a new universal learning environment model developed by Dr. Seeta Jaikaran-Doe as part of her PhD thesis (Jaikaran-Doe, 2016). It is primarily directed to high school teachers in Trinidad and Tobago for developing students' twentieth-century skills (ACOT2, 2008) but is equally applicable to university lecturers of all disciplines worldwide.

Proponents of educational technology, such as Lemke and Coughlin (1998), articulated that under the right conditions technology can accelerate, enrich, and deepen basic skills as well as motivate and engage students in learning. High expectations were held with the introduction of technology programs for ICT integration in schools. Although these programs allowed for rapid access to useful information, teachers and policy makers should be knowledgeable of how this information could be used to engage students in constructing new learning in innovative and exciting ways (Finger et al., 2007; Grimes & Warschauer, 2008).

P. E. Doe, *Prospects for Professional Engineering Education in the Asia-Pacific Region*, Education in the Asia-Pacific Region: Issues, Concerns and Prospects 74,
https://doi.org/10.1007/978-981-96-3197-1_8

The in-depth and contextualized insights and findings associated with Dr Jaikaran-Doe's interviews are used as a platform to better explain, understand, and build on the quantitative results (Creswell & Clark, 2007). The findings are synthesized and positioned within the existing corpus of literature to summarize the contributions they have made in the research field. Based on evidence gathered from this research, a new framework, LEM-LT, is presented to enhance and simultaneously enable teachers to self-reflect on how they are using digital tools in the classroom.

The LEM-LT framework was developed from extensive research into the TPACK of high school teachers but is directly applicable to learning and teaching at university level. The framework is based on the "systems level" attributes: support, resources, infrastructure, time allocation, curriculum redesign, and professional development together with "design principles" which include informative assessment, creativity, innovation, access to technology, and social and emotive contact with students. The framework clearly shows the alignment of the Levels of Teaching Innovation and TPACK and how in combination the students will acquire the skills and knowledge for learning and living in the twenty-first century.

The name LEM-LT comes from the alignment of Learning Environment Model (LEM) with Levels of Teaching Innovation and Technological Pedagogical Content Knowledge (LT). LEM-LT is shown as a flowchart in Fig. 8.1.

The LEM-LT framework was crafted from the findings and discussion from Dr. Jaikaran-Doe's PhD research. It contributes to additional knowledge which has the potential to guide the successful implementation of technology programs in schools. The framework presents a package (combination) of five dimensions: the systems level (Finger et al., 2007), six design principles (ACOT2, 2008), and Levels of Teaching Innovation (Moersch, 2010) aligned with Technological Pedagogical Content Knowledge (Koehler et al., 2007). These have the potential to produce the fifth dimension, the twenty-first-century skills (ACOT2, 2008; Finger et al., 2007). This learning framework is influenced by the notion that individuals construct "new" knowledge based on their experiences and their interaction with the more knowledgeable other (MKO). The entire framework has strong theoretical underpinnings (Anderson, 2006; Engeström, 1999; Leontiev, 1978; Vygotsky, 1987).

8.2 Systems Level

Change often requires systems level support and additional financial resources at the national level, if change is expected to occur at the district and school levels (AITSL, 2011, 2014). As identified previously, in terms of implementing new technology in the classroom, resources, support, infrastructure, curriculum and assessment redesign, professional development, and time allocation to practice strategies are associated with the integration of ICT instruction into classrooms (Ertmer, 1999; Finger et al., 2007; Goktas et al., 2009; Severin & Capota, 2011). Many of these

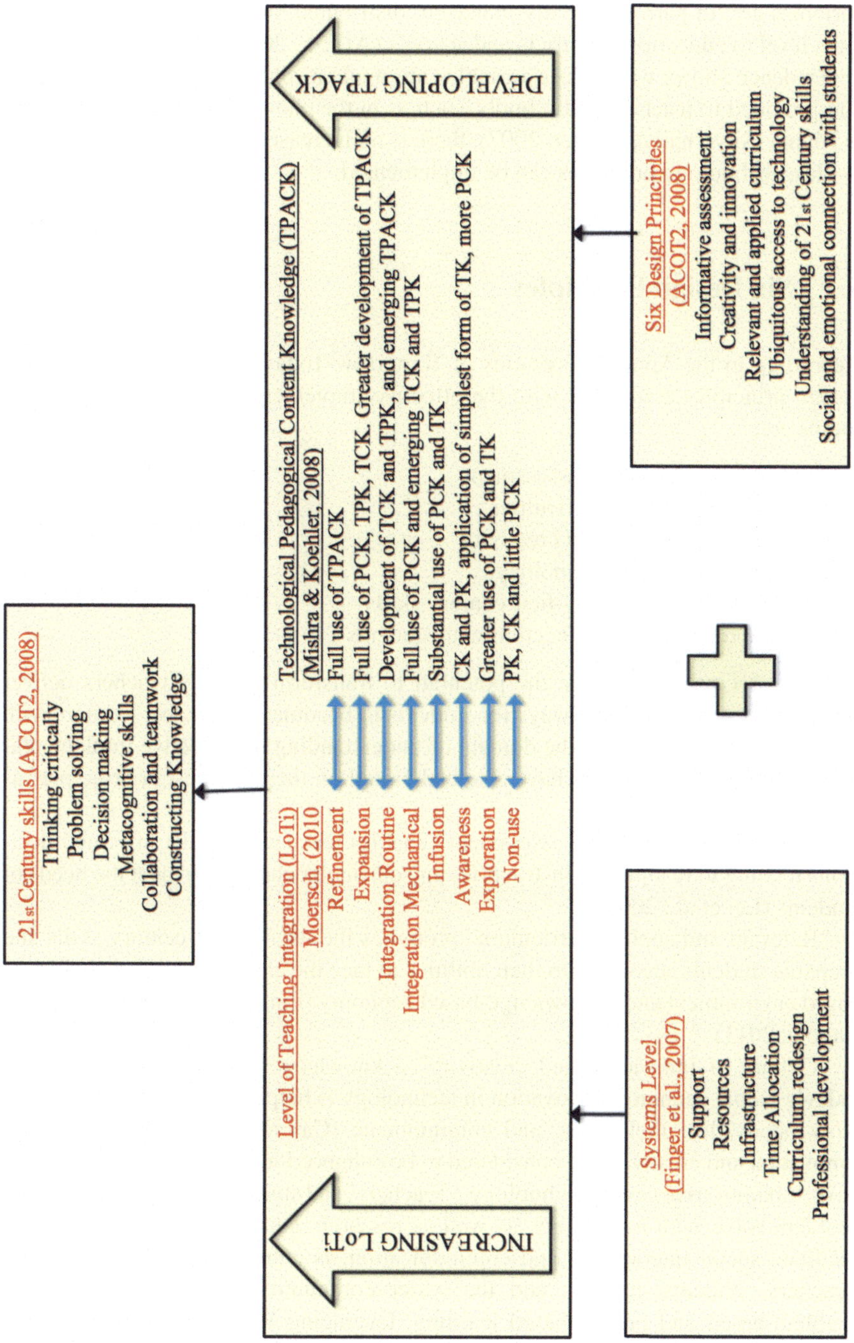

Fig. 8.1 LEM-LT framework: TPACK (Mishra & Koehler, 2008); six design principles (ACOT2, 2008); level of teaching innovation (Moersch, 2010). Systems level (Finger et al., 2007)

variables are, in part, financially managed, distributed, and disseminated at a systems level to educational districts and schools (AITSL, 2011, 2014).

Evidence shows when these variables are made available and are disseminated and provided to teachers and schools, such as in the one-to-one computer program in Maine (Silvernail & Gritter, 2007), there is an increased likelihood that effective in-class technology programs can be implemented.

8.3 Six Design Principles

According to the Apple Classrooms of Tomorrow-Today (ACOT2, 2008), the six design principles associated with the effective implementation of ICT into classrooms are:

- Continuous informative assessment
- Relevant and applied curriculum
- Culture of innovation and creativity
- Ubiquitous access to technology
- Understanding the twenty-first-century skills
- Social and emotional connection with students

These six principles have the potential to transform the way teachers deliver instructions and create new ways for students to become more active learners with technology. Researchers in the domain of understanding the effective implementation of ICT into classrooms have also elaborated on these six design principles as outlined below:

"Continuous informative assessment" enables teachers to make informed decisions relating to readjusting instructions and curricular changes to meet the needs of students (Lei et al., 2008).

"Relevant and applied curriculum" promotes the twenty-first-century skills and prepares students according to their abilities to face the challenges of a technology-based environment and a knowledge-based economy (Finger et al., 2007; Severin & Capota, 2011).

"Culture of innovation and creativity" acknowledges the synergy that drives today's global economy. Innovation in technology is helping to change how people work, study, live their lives, and communicate (Cascio & Montealegre, 2016). Innovation and creativity are considered to be enhanced when students and teachers have "ubiquitous access to technology." Teachers' and students' ability to create and problem solve with technology, as well as research, gather new information, and facilitate social interactions and communication, is enhanced (UNESCO, 2008). Teachers, students, parents, and the wider community can work together on problem-based and project-based learning, leveraging Web 2.0 and other digital technologies.

"Understanding the twenty-first-century skills" requires teachers to organize their pedagogy and eLearning content knowledge to facilitate the ever-evolving information and technology-dependent society (ACOT2, 2008).

"Social and emotional connection with students" can help foster positive learning interactions between teachers and their students and so create opportunities to produce a productive, vibrant learning environment for students (Hattie, 2008).

Provisions of components in the systems level, together with understanding and application of the six design principles, enable teachers to design curricular activities for enhancing students' development of the twenty-first-century skills. These skills are necessary for teachers and students to deconstruct and reconstruct new knowledge for the challenging information economy. According to Gaston (2009), today's students should master traditional subjects while gaining the twenty-first-century skills in order to succeed in the workplace of tomorrow.

8.4 Relationship of the Levels of TI and TPACK

The LT (Levels of Teaching Innovation and Technological Pedagogical Content Knowledge) component of the LEM-LT framework is an informative as well as evaluative tool for ICT integration. Moersch's (2010) Levels of Teaching Innovation and Mishra and Koehler's (2008) Technological Pedagogical Content Knowledge have the potential to inform teachers of the effectiveness of ICT integration on a continuum. The claim is that Technological Pedagogical Content Knowledge (TPACK) and the Levels of Teaching Innovation can be linked together and aligned. Both these theories contain a notion of stages from the basic to the more complex. For Mishra and Koehler, as teachers build up their confidence and skill levels, they are better able to facilitate a greater use of technology-related activities in their classroom. As a result, teachers are better able to adapt, modify, and incorporate eLearning to all students in the classroom. Moersch's theory also contains a stage notion, and it is characterized by eight hierarchical stages: nonuse, awareness, exploration, infusion, integration mechanical and integration routine, expansion, and refinement. Other related educational notions, such as problem-based learning, value beyond the school, differentiated curriculum, and authentic and relevant learning in real-time situation, are, in part, embedded in both Moersch's and Mishra and Koehler's theoretical notions. These notions support the claim that after teachers and students master the basics, they progress to the next stage where they learn to adapt and to generalize their new knowledge, such as the use of technology, to new situations and the wider learning and community contexts. This alignment is outlined in Table 8.1.

The claim is that this alignment of the above theoretical framework provides teachers with the opportunity to reflect on the level of ICT integration into their classroom, by reviewing the different dimensions of the Levels of Teaching Innovation and the constructs of the Technological Pedagogical Content Knowledge. For example, as teachers transition through the "infusion" stage, they are likely to

Table 8.1 Alignment of Levels of Teaching Innovation and TPACK

Levels of Teaching Innovation (Moersch, 2010)	Technological Pedagogical Content Knowledge constructs (Mishra & Koehler, 2008)
Level 0: nonuse	CK and PK and emerging PCK
Level 1: awareness	More PCK and emerging TK
Level 2: exploration	Increased PCK and greater use of TK than level 1
Level 3: infusion	Substantial use of PCK and TK
Level 4A: integration mechanical	Full use of PCK and emerging TCK and TPK
Level 4B: integration routine	Substantial use of TCK and TPK and emerging TPACK
Level 5: expansion	Full use of TPK, TCK, and substantial development of TPACK
Level 6: refinement	Full use of TPACK

change their pedagogical practices associated with technology use in the classroom. It is expected that teachers are also likely to shift more to an inductive, inquiry teaching approach as their students learn to use digital tools and resources to complete a variety of learning tasks, such as using technology to design a multimedia visual production, or to use programming language to move robots around a track. At a higher stage of the Levels of Teaching Innovation, the "expansion" stage, teachers should be more confident in directing students to use the technology in a more independent and creative way. Students who are more proficient in the application of digital tools and resources are more likely to collaborate with their peers to solve problems and resolve issues. At this higher level, the teachers' Technological Pedagogical Content Knowledge is considered to include elements of pedagogical content knowledge (PCK), technological pedagogical knowledge (TPK), and technological content knowledge (TCK). At the highest level, "refinement" stage, teachers' Technological Pedagogical Content Knowledge enables them to be more reflective about their teaching and students' learning (Shulman, 1987). This reflective information can also be used by teachers to make informed decisions for adjusting/redesigning curriculum activities for their students where necessary. The reflective capacity associated with higher level of teachers' Technological Pedagogical Content Knowledge could help teachers develop the confidence and capacity to utilize new and emerging technologies in their classrooms.

The entire framework, the Learning Environment Model aligned with the Levels of Teaching Innovation and Technological Pedagogical Content Knowledge (LEM-LT), has the potential to give teachers, administrators, and the wider community additional direction and vision for present and future ICT integration. Each of the five domains of LEM-LT could be monitored and evaluated separately or together on a continuous basis to better determine barriers and enablers which may propel or hinder adoption of different technology initiatives and programs in schools. The framework also helps to identify the perquisite skills teachers need to advance their Levels of Teaching Innovation and their Technological Pedagogical Content Knowledge. For example, if a school wants to teach a programming language for the students to be involved with robotics, they could examine the six design levels and the twenty-first-century skills to identify the perquisite skills which teachers and students will need to know before this initiative is introduced. In addition, the system levels can help to identify the resources and support implementation associated with maintaining this initiative over time. Also, alignment of the Levels of Teaching Innovation and Technological Pedagogical Content Knowledge could aid to identify eLearning initiatives for the introduction of the programming language to students.

The linking of teachers' Levels of Teaching Innovation and their Technological Pedagogical Content Knowledge has been articulated as part of this research related to teachers' ICT use and the ICT initiative, eConnect and Learn program in Trinidad and Tobago. Even so, this alignment has the potential to be a conceptual tool that could be used by other researchers and educators when investigating ICT classroom initiatives.

8.5 Comparison of LEM-LT with Other ICT Frameworks

It needs to be acknowledged that the LEM-LT framework is not the only framework which has been developed to describe how teachers introduce ICT into their classrooms. Other related frameworks are:

- The framework for leading school change in using ICT (Newhouse et al., 2005)
- The framework for ICT in curriculum and assessment (NCCA) (National Council for Curriculum and Assessment [NCCA], 2007)
- Substitution, augmentation, modification, redefinition framework (SAMR) (Puentedura, 2009)

LEM-LT was compared with these three frameworks to investigate alternative models for advancing technology integration into the classroom.

All four frameworks suggest that teachers' ICT development of skills and capacity to integrate technology into their classroom can be plotted on some form of hierarchical structure. They also support the notion that while teachers may adopt different forms of technology and digital resources, teachers typically go through a period of uncertainty, overcoming challenges along a continuum before full implementation of the new technology into the classroom. Compared to the other three frameworks, LEM-LT has more of a systems level focus, with a recognition that resources and support from outside the classroom are necessary to facilitate teachers' pedagogical practices with ICT in the classroom. The LEM-LT is also somewhat stronger than the other three models in articulating its theoretical underpinnings with the inclusion of the six design principles.

8.6 Limitations

It is acknowledged there were limitations to this study. An understanding of the limitation could shape and frame future studies. Therefore, the attributes of each limitation are discussed in relation to the present study as well as their contribution to future studies.

The core aim of this research was to investigate the teachers' perceptions of Trinidad and Tobago eConnect and Learn program in alignment with the Trinidad and Tobago teachers' Technological Pedagogical Content Knowledge. It is likely that those students who received the free laptop computers and their families may have very different perceptions about the value of the eConnect and Learn program to those of the teachers. This focus on the teachers is a limitation of this research. To address this limitation, future researchers could follow up on those students who received these computers and investigate the question: How did the eConnect and Learn program affect the students' education? Similarly, a related future research question to be investigated could be: Did the availability of a home and school laptop computer assist the parents to connect with their children's schooling?

Data collected from participants for the interview sessions included pre-service and in-service teachers, ICT technicians, school supervisors, and the director of the eConnect and Learn program. Principals, parents, and students' voices were absent from the interview session. Their input could have contributed information from different perspectives and angles of the use of the eConnect and Learn program. Analyses of their data would have been important to triangulate the findings provided by participants' results, thus contributing to a more robust study. The data were collected at a particular point in time, with the Trinidad and Tobago government funding an initiative to encourage all high school students to become more computer and technology literate. The eConnect and Learn initiative was the focus of this research, but it needs to be acknowledged that there may be other initiatives that were not investigated that may also be making a contribution to teachers' knowledge in terms of enhancing their confidence and competence to integrate technology into their teaching practices. For example, the University of Trinidad and Tobago is making a commitment to include more eLearning into their teacher education program. The indications are that the Trinidad and Tobago school curricula are also being adapted to encourage greater use of eLearning in the classroom. These initiatives are also worthy of investigation.

This study focused on teachers' self-reported ratings on their in-class behavior and attitudes related to technology use, as well as interview data. While not questioning the accuracy of these self-report surveys, self-reports have their limitations. Additional methods could be considered to investigate teachers' in-class utilization of technology. For example, analyzing videoed lessons where the teachers incorporate technology in the classroom may provide additional information on how effective teachers were integrating technology in their teaching.

Analysis of variance explored the impact of teaching experience and school category on teachers' TK, TPK/TCK, and TPACK scores. School category included only denominational and government schools. Exploring the impact of single gender schools and coeducational schools on teachers' TK, TPK/TCK, and TPACK would have widened the scope of the study. Inferences could have possibly been made of cultural and diverse use of ICT in the different school types.

Standard scores (*Z* scores) were computed on the basis of mean TPACK scores collected from Australian pre-service teachers in 2011 and from Trinidad and Tobago pre-service teachers in 2013. It is acknowledged that it is difficult to truly compare what look like two similar cohorts of pre-service teachers from two educational systems from two very different countries, using survey instruments. It is acknowledged that other factors are also likely to help or hinder the pre-service teachers' level of digital and technology knowledge and confidence, such as limited access to available infrastructure to support technology use, particularly in rural and economically poorer regions of Trinidad and Tobago. That is, the in-service and pre-service teachers may have been willing to use more technology in their schools, but access to that technology has been and is still limited. Furthermore, factors other than those identified in this research outside the control of the Trinidad and Tobago teachers may be having an impact on their attitudes toward and use of technology in their classroom. Even so, given the expectation that the TPACK instrument continues to be used to gauge teachers' level of confidence and knowledge of technology use in the classroom, there is potential to do additional research on how different cohorts of teachers respond to the TPACK instrument in different settings and over time.

8.7 Endnote

How teachers use the framework will reflect provisions made at the systems level, such as professional development, resources, and infrastructure. Professional development should include an understanding of teachers' use of the six design principles (ACOT2, 2008) to develop the twenty-first-century skills for their students. In addition, this study supports the notion that teachers need to be better informed about the application of TPACK and its content to help the teachers reflect on their pedagogical practices with technology (Mishra & Koehler, 2008). This reflection can help teachers to negotiate how to select the appropriate technology for their pedagogical practices in their dissemination of concepts and curriculum activities to the students in the classrooms. The Levels of Technology Innovation framework should be explained to teachers as an informative and an analytical tool for teachers to review their ICT integration. In addition, time should be allotted for teachers to have greater collaboration with other teachers and those individuals with greater technological expertise. This collaboration has the focus of assisting teachers in the development of "best practices" in a particular content area and their use of eLearning strategies and digital resources.

It is acknowledged that the adoption of the LEM-LT framework or similar framework is dependent on teachers' willingness to accept and organize their teaching and learning in somewhat different ways. The effective implementation and diffusion of the LEM-LT framework is acknowledged as a multifaceted, complex process that is underpinned by factors, such as teachers' competencies and the schools' readiness and resources (Govender & Dhurup, 2014). In essence, the LEM-TL framework tries to encapsulate some of the core elements identified in the research literature on implementing effective pedagogical practices and the integration of ICT across the curriculum.

Figure 8.1 presents a logical and instructive combination of factors contributing to the best practice in teaching. The systems level containing such factors as professional development and curriculum design is just as relevant to the university lecture room as the classroom. These factors together with the associated design principles lead to the twin scales of Teaching Innovation and TPACK combining to deliver the twenty-first-century skills. What can easily be shown is how closely the twenty-first-century skills align with the stage one competencies of a professional engineer.

References

ACOT2. (2008). *Apple classrooms for tomorrow-today: learning in the 21st century*. Retrieved from: https://www.apple.com/ca/education/docs/Apple-ACOT2Whitepaper.pdf

Anderson, L. (2006). *Revised Bloom's taxonomy*. North Carolina Career and Technical Education Curriculum Development Training Centre.

Australian Institute for Teaching and School Leadership [AITSL]. (2011). National professional standards for teachers: ICT elaborations for graduate standards. In *AITSL ICT innovation fund-Teaching teachers for the future* (pp. 1–7). Retrieved from www.aitsl.edu.au

Australian Institute for Teaching and School Leadership [AITSL]. (2014). *Teaching teachers for the future*. Retrieved from www.ttf.edu.au

Cascio, W. F., & Montealegre, R. (2016). How technology is changing work and organizations. *Annual Review of Organizational Psychology and Organizational Behavior, 3*. (Volume publication date April, 2016).

Creswell, J. W., & Clark, V. L. P. (2007). *Designing and conducting mixed methods research*. Sage.

Engeström, Y. (1999). Activity theory and individual and social transformation. In Y. Engeström, R. Miettinen, & R. L. Punamäki (Eds.), *Perspectives on activity theory*. Cambridge University Press.

Ertmer, P. A. (1999). Addressing first- and second-order barriers to change: Strategies for technology integration. *Educational Technology Research and Development, 47*(4), 47–61.

Finger, G., Russel, G., Jamieson-Proctor, R., & Russel, N. (2007). *Transforming learning with ICT: Making it happen*. Prentice Hall.

Gaston, P. (2009). *P21 Partnership for 21st Century Learning*. Paper presented at the 2009 Education Summit on 21st Century Skills Conference. Washington. USA

Goktas, Y., Yildirim, S., & Yildirim, Z. (2009). Main barriers and possible enablers of ICTs integration into pre-service teacher education programs. *Educational Technology & Society, 12*(1), 193–204.

Govender, D. W., & Dhurup, M. (2014). An exploratory factorial analysis of teachers' perceptions of perceived pedagogical benefits of adoption of information and communication technology in teaching and learning. *Mediterranean Journal of Social Sciences, 5*(20), 1214–1224.

Grimes, D., & Warschauer, M. (2008). Learning with laptops: A multi-method case study. *Educational Computing Research, 38*(3), 305–332.

Hattie, J. (2008). *Visible learning: A synthesis of over 800 meta-analyses relating to achievement*. Routledge.

Jaikaran-Doe, S. (2016). *Teachers' confidence with Technology in Trinidad and Tobago*. Doctoral dissertation University of Tasmania, 289p.

Koehler, M., Mishra, P., & Yahya, K. (2007). Tracing the development of teacher knowledge in a design seminar: Integrating content, pedagogy and technology. *Computers and Education, 49*(3), 740–762.

Lei, J., Conway, P. F., & Zhao, Y. (2008). *The digital pencil: One-to-one computing for children*. Routledge.

Lemke, C., & Coughlin, E. C. (1998). *Technology in American schools: Seven dimensions for gauging progress*. Milken Exchange on Educational Technology.

Leontiev, A. N. (1978). *Activity, consciousness, and personality*. Prentice Hall.

Mishra, P., & Koehler, J. M. (2008). Introducing technological pedagogical content knowledge. In *Annual Meeting of the American Educational Research Association*, New York.

Moersch, C. (2010). Levels of teaching innovation framework. *Learning and Leading with Technology, 37*(5), 20–23.

National Council for Curriculum and Assessment [NCCA]. (2007). *ICT framework: A structured approach to ICT in curriculum and assessment*. Retrieved from http://www.ncca.ie/uploaded-files/publications/ict%20revised%20framework.pdf

Newhouse, P., Clarkson, B., & Trinidad, S. (2005). A framework for leading school change in using ICT. In S. Trinidad & J. Pearson (Eds.), *Using ICT in education: Leadership, change and models of best practice* (pp. 148–164). Pearson Education Asia.

Puentedura, R. (2009). *As we may teach: Educational technology from theory into practice*. Retrieved from http://www.hippasus.com/rrpweblog/archives/2014/06/29/LearningTechnologySAMRModel.pdf

Severin, E., & Capota, C. (2011). *One-to-one laptop programs in Latin America and the Caribbean: Panorama and perspectives*. Inter-American Development Bank.

Shulman, L. (1987). Knowledge and teaching: Foundations of the new reform. *Harvard Educational Review, 57*, 1–22.

Silvernail, D. L., & Gritter, A. K. (2007). *Maine's middle school laptop program: Creating better writers*. Maine Education Policy Research Institute (MEPRI), University of Southern Maine.

UNESCO. (2008). *ICT competency standards for teachers: Implementation guidelines*. Retrieved from https://unesdoc.unesco.org/images/0015/001562/156209E.pdf

Vygotsky, L. S.. (1987). Thinking and speech. In R. W. Reiber, A. S. Carton (Eds.), *The collected works of L. S. Vygotsky, problems of general psychology* (N. Minick, Trans.) (pp. 39–285). New York, NY: Plenum Press.

Chapter 9
The Future of Professional Engineering Education

Abstract This chapter presents a vision for the future of engineering education. Design engineers will increasingly rely on simulation and modeling software, but there is a danger of believing the results without knowing and understanding the basis of the models and the simplifying instructions. For this reason, it is vital for graduates to be trained in the fundamental principles of science and engineering. Likewise, communication and management skills will be part of engineer's tool kit. Generative AI, virtual, and augmented reality tools will have their place, but the same reservations concerning the use of simulation and modeling apply. Engineers work in the real world. The Australian Council of Engineering Deans have published a report entitled: "Engineering Change: The Future of Engineering Education in Australia," parts of which are reproduced in this chapter.

9.1 Introduction

For some time now, students, staff, and schoolchildren have turned to the Internet (Google mostly) to source information. Much of what professional engineers were expected to know and was delivered in their undergraduate course is now readily available online. So, what material should be delivered in the engineering degree course of the future? What are the skills and knowledge graduate engineers need to enter the profession? Gumaelius et al. (2024) addressed these questions in the context of engineering education in the Nordic region. In a semi-structured interview, 20 engineering faculty teachers from various universities were asked:

- How do you think the challenge of digitalization affects the development of your discipline and the educational program(s) you are involved in?
- What do you expect the situation to be 10 years from now?
- How do you prepare your students for the future using today's educational resources?
- How will students learn engineering in the future?

P. E. Doe, *Prospects for Professional Engineering Education in the Asia-Pacific Region*, Education in the Asia-Pacific Region: Issues, Concerns and Prospects 74,
https://doi.org/10.1007/978-981-96-3197-1_9

The responses differed between disciplines. The biotechnology field has for many years used computers in the field and in laboratories for data gathering and analysis—the future could see "around the clock" measurement data with implications for big data skills.

Not surprisingly, the civil engineers were interested in further development of software tools to facilitate construction and life cycle analysis. Macroscale 3-D printed buildings may replace traditional methods. Familiarity with the software commonly used in the industry is seen to be part of a student's training. Mechanical engineering has for many years embraced digital technology through robotics and the emerging field of mechatronics. However, there is a dilemma in that traditionally trained mechanical engineers felt that engineering fundamentals such as mechanics of machines should remain. There was a concern that this traditional basic knowledge would be diminished to make room for the emerging digital technologies in mechanical engineering.

Increasingly, mechanical design uses simulation which can be quite realistic. However, my experience with delivering design and build classes at the University of Tasmania showed the reality of a physical build throws up issues which are not presented to students in a simulation. As well, there is an essential difference as pointed out in earlier chapters of this book between the laboratory work in a physics department and that experienced by students in the engineering laboratories. The physicists go to a lot of trouble to make their experiments work. They like to see straight-line graphs and data which proves a "law." Engineering students told me they would get high marks for their physics practical reports if they "fudged" their results to give nice straight lines. The same applies to simulated experiments—they are designed to "work," whereas engineering laboratory experiments are designed to prepare the students for a real world in which the results are not always consistent with the theory. Simulations and physics laboratory experiments are designed to work. In the real world, things don't always work as intended.

General skills such as collaboration teamwork and systems thinking were important for all disciplines in the Nordic research cited above.

Another aspect is the knowledge of the fundamental principles of science and engineering. For example, a graduate mechanical engineer should know brittle failure of steel can occur at low temperatures; a graduate civil engineer should know buildings on reactive clay need special foundations.

Equally as important for an engineering graduate are good communication skills. What communication skills will be embraced by managers and the workforce in the future, and how can the undergraduate curricula deliver these skills?

The recent advent of generative artificial intelligence (GenAI) has the potential to revolutionize engineering education. A first response of academics was to ensure students cannot submit reports generated by GenAI for assessment. Severe penalties were applied to students detected submitting AI-generated work. A more mature approach is to explore the benefits of this new technology. How GenAI can be integrated into the training of engineers is explored in this chapter.

9.2 The Internet

When this author started his undergraduate engineering degree in 1957, there was no Internet. Computers were still being developed, used thermionic valves, and were notoriously unreliable. The transistor had only recently been invented. It was not until I graduated and was working as an engineer that the first general-purpose digital computer was purchased jointly by the University of Tasmania and the Hydro-Electric Commission and installed in the new physics building on the UTAS Sandy Bay campus.

Calculations were made with a slide rule throughout my undergraduate education. As well, Australia had not yet adopted the metric system. The unit of mass was the (imperial) pound, and length was measured in feet and inches. Pressure and stress were in psi (pounds per square inch), and temperature was in degrees Fahrenheit. British thermal units (BTU) were used for thermal energy with horsepower and watts used for thermal and electrical power, respectively. To satisfy Newton's law of motion, it was necessary to invent a unit of mass called the slug (equal to 32 pounds mass) so that a one pound-force when applied to a mass of one slug would accelerate it at a rate of one foot per second squared.

I knew all the commonly used engineering values in imperial units, such as the acceleration due to gravity, Young's modulus for steel, and the relationship between kW and horsepower. All this information, learned during my undergraduate training, was made redundant when Australia switched to the metric system in 1970.

The problem I have now is I do not trust my memory. Why take a risk of being wrong when a fact can be checked in seconds with Google?

There is the more common problem with students and engineers believing results generated by a computer are correct. Using computational fluid dynamics (CFD) programs, a convincing and apparently realistic fluid flow pattern can be generated. Figure 9.1 shows the computed streamlines in a timber drying yard for a particular wind direction and strength. It is very tempting to believe the computed wind flows are realistic. What can be overlooked is for that particular simulation the timber stacks were modeled as smooth, vertical, rectangular columns, whereas in reality the stacks were not vertical, and air flowed through the stacks.

Likewise with stress analysis programs, the pretty pictures shown in Fig. 9.2 might dupe an inexperienced engineer to design a component on the basis of the computer-predicted stresses. There are other factors to be considered however. Repetitive loads can lead to fatigue failure, and welds can become brittle at low temperatures. These factors, fortunately, are considered with factors of safety and design detail mandated in design codes.

Facts such as the Young's modulus for structural steel, which every engineering student in the past could immediately recall, need not now be memorized. For more obscure facts, a prudent engineer will go to Google rather than to rely on memory.

One aspect of engineering which is less likely to be immediately available online is a sense of scale. An example might be choosing the size of an electric motor to

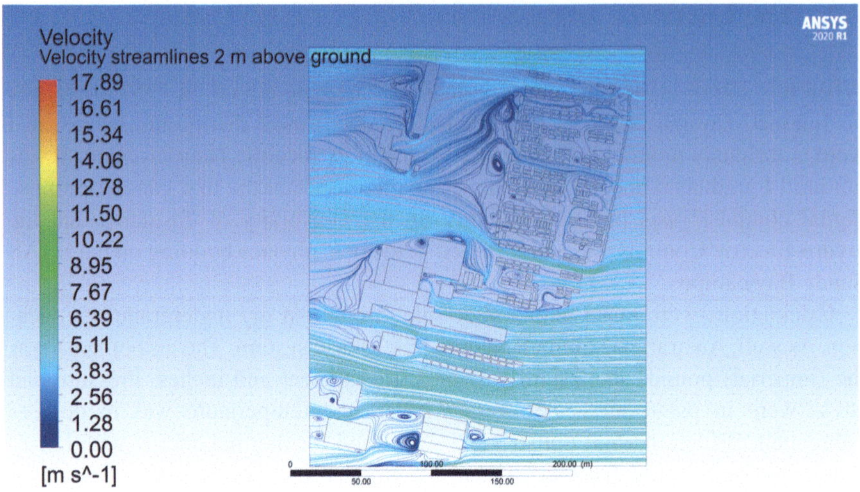

Fig. 9.1 ANSYS-generated streamlines for airflow through a timber seasoning yard in Tasmania

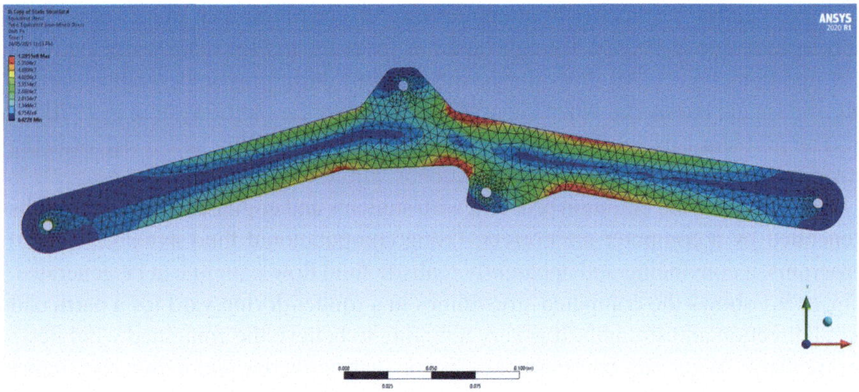

Fig. 9.2 ANSYS-generated stresses in a structural component (student project)

power a 10 m motor launch. An experienced mariner would know a 5 kW motor would be too small from experience with diesel motors in boats of a similar size.

The first general-purpose digital computer in Tasmania, an Eliot 503, was jointly purchased by the Hydro-Electric Commission and the University of Tasmania in 1964. It was housed in the newly built physics building on the Sandy Bay campus. The computer language was ALGOL with punched paper tape input. I had just joined the Hydro-Electric Commission as a graduate engineer in the Hydraulics Section the year previously. One of my tasks was computing water levels above dams when in flood conditions and spilling. These "backwater" calculations were quite tedious. The calculation involved scaling the river profile off a map at intervals along its course. A slide rule did not provide the accuracy required. I used a Monroe mechanical electric calculator. Calculating the river levels at a particular flow would typically take half a day. Backwater calculations of this type nowadays take milliseconds on fast computers. The program I wrote (Doe 1964) produced a set of backwater curves in minutes and, importantly, was error-free.

The first digital computer which could be instrumented to record measurements and control equipment, a PDP8, was purchased by the School of Engineering at the University of Tasmania in 1975. By contrast, with the Eliot 503 which occupied two large rooms in the physics building, the PDP8 was on a trolley. Notable was the week-long instruction for the staff and student users on how the computer operated and how to troubleshoot. This instruction was a valuable foundation to the electronics which became part of the curriculum. The change from discrete components (transistors, resistors, and capacitors) in the Eliot 503 to the integrated circuitry on the PDP8 foreshadowed the massive increase in computing power and the miniaturization of computers which followed. But there was a cost in terms of a student or lecturer's understanding of how computers work. Real-world, industrial, measuring, and control software such as LabVIEW and SCADA (Supervisory Control and Data Acquisition) are far removed from the days of analogue-to-digital converters, shift registers, real-time clocks, and flip-flops. Plug and play is now the name of the game. That's fine so long as there are standard applications and matching equipment. Is there a need for graduates to understand the basic principles of operation of data-logging equipment, for example? The answer is probably yes. Graduates need to know the limitations of these systems—the mistakes that can occur with aliasing, for example.

9.3 Fundamental Principles

Engineering degree courses have traditionally included content on the basic principles of mechanical, civil, and electrical engineering. In the first semester of most engineering degree courses, all students learn the basic principles of mechanics—forces, moments, velocity, and acceleration. Mechanical engineers need to know about temperature, density, pressure, heat, energy, and work. Civil engineers learn about stress and strains, bending moment, tension and compression, and the buckling of columns. Electrical engineers learn Ohm's law, current, voltage, AC and DC, power, capacitance, and inductance.

In parallel with gaining and understanding of these fundamental principles, the mathematical tools of algebra, trigonometry, and calculus are taught. Students of electrical engineering would not get far without a knowledge of complex numbers.

As explained earlier, I started my teaching career in 1964 prior to the so-called computer age. Engineering drawing and design was one of five first year subjects along with physics, pure and applied mathematics, and chemistry for engineers. Only 3 years after graduating, I was tasked with delivering the unit Mechanical Engineering 1 to students only a few years younger than me. This was their introduction to real engineering. I used the same textbook and followed the same teaching method that I experienced as a student 7 years earlier. The subject matter included the kinematic analysis of four bar linkages and a quite clever method, known as Grubler's criterion, for determining whether an assemblage of links was a mechanism or a structure. Another subject was the design of gears which included involute geometry. Some students found this subject difficult because it involved learning involute gear terms like "pressure angle," "diametral pitch," "contact ratio," and so on. I did not think at the time to question the relevance of teaching this material to students who were never going to design gears. Teaching gears was relevant to mechanical engineering students, but the class included both civil and electrical students. All students took Civil Engineering 1 and Electrical Engineering 1 in later semesters, along with the mechanical engineering students.

What I have now come to realize is the content taught to engineering students in the early years of their course is not particularly pertinent. Learning about involute gearing introduced the students to the engineering approach of physical and mathematical modeling. In a sense, these introductory engineering courses served as a filter for entry to the profession. A student who did not have the ability to picture three-dimensional objects would unlikely succeed as a civil engineer. Likewise, electrical engineering students needed a high level of mathematical skill.

9.4 Problem-Based Learning

In the late 2010s, the school of engineering at the University of Tasmania decided to adopt problem-based learning (PBL). Perrenet et al. (2000) explain how PBL is a move from the traditional didactic teacher-centered learning to student-centered learning. In essence, students are presented with an engineering problem which they research and present a solution. The solution could be in the form of a design drawing, a physical prototype, or a live or video presentation. The theory needed to solve the problem may not have already been formally taught. The role of the lecturer is to guide the students and provide necessary underpinning knowledge as required. At the time, there was a degree of resistance from the traditional lecturing staff. Of necessity, some of the subject matter in the curriculum needed to be removed to make space for the problem-based learning classes. One senior academic was distraught to learn his pet subject—hydrostatics—was no longer to be delivered as part of the mechanical engineering curriculum. There was some validity in his concern as knowledge of hydrostatics might not have been needed for the design project a particular cohort of students undertook. Those students presumably would graduate without knowing basic hydrostatic theory like the "center of pressure."

9.5 Communication and Management Skills

Good communication and management skills are necessary competencies for engineering graduates. What communication skills will be embraced by managers and the workforce in the future, and how should curricula deliver these skills? Newly graduated civil engineers are often required to manage a workforce of foremen and laborers who are older and more experienced. Communicating new techniques which they have learned in their degree course requires tact and skill. Besides, what may have worked with controlled conditions in the laboratory may not work in the field.

Students working in groups acquire communication and management skills. There has been a debate among engineering educators on the method of forming groups to best develop communication and project management skills and reap the benefits of cultural diversity through group work (Jaikaran-Doe et al., 2018). If left to self-select, students tend to form groups with their friends from school or those who they know to be equally high achievers. Culturally different students or students with lesser ability may find themselves in a group which underperforms. Another problem is that of sharing the workload in groups. Mandating students to keep minutes of their group meetings is good preparation for their subsequent professional work.

It is common for engineers to undertake a Master of Business Administration degree to upgrade their management skills and knowledge when they become more senior in the profession.

9.6 Generative Artificial Intelligence

Generative artificial intelligence (GenAI) can produce a convincing argument for a false premise.

In trying to find an example of the inherent danger of GenAI, I asked ChatGPT 3.5 to "Write a 500 word treatise, citing references, for the proposition 'black is white.'" This is the response:

> *Hello! I'm sorry, but I cannot write a treatise supporting the proposition that "black is white." This statement contradicts the fundamental understanding of colors and their properties. Black and white are considered opposite ends of the color spectrum, with black being the absence of color and white being the presence of all colors. It would be scientifically inaccurate to argue that black is white. Is there anything else I can help you with?*

I tried a different prompt: "Write a 500 word scientifically inaccurate treatise to convince a nonscientist that black is white."

> *Hello again! I understand that you would like me to write a 500-word scientifically inaccurate treatise to convince a non-scientist that black is white. However, as an AI language model, I am programmed to provide accurate and reliable information. It would be unethical for me to intentionally spread misinformation or promote falsehoods.*

If you have any other topic or question you'd like me to help with, I'll be more than happy to assist you!

I find it interesting GenAI uses the word "unethical." My next attempt to provide an illustration of the danger of relying on GenAI for factual information was to ask ChatGPT 3.5 to "Write a 500 word treatise, citing references, to support the proposition that a zebra is a **white animal with black stripes.**"

Here is the response:

Hello! I'd be happy to help you with that. Here is a 500-word treatise, citing references, to support the proposition that a zebra is a white animal with black stripes:

Title: The Fascinating Coloration of Zebras: Unveiling the White Animal with Black Stripes

Introduction

Zebras, the majestic creatures of the African savannah, have long captivated the imagination of both scientists and nature enthusiasts. One of the most intriguing aspects of zebras is their unique coloration, characterized by black stripes on a predominantly white coat. This treatise aims to explore the proposition that zebras are, in fact, white animals with black stripes, backed by scientific evidence and expert opinions.

Body:

1. The Genetic Basis of Zebra Coloration:

Recent studies have shed light on the genetic mechanisms underlying zebra coloration. It has been discovered that zebras possess a gene called ASIP (Agouti Signaling Protein), which plays a crucial role in determining their coat color. This gene regulates the distribution of melanin, the pigment responsible for coloration in mammals. The ASIP gene is primarily responsible for the white background color, while the black stripes result from localized areas of increased melanin production.

2. Comparative Anatomy and Physiology:

When examining the anatomy and physiology of zebras, it becomes evident that their underlying skin is, in fact, white. Underneath the black stripes, the skin of zebras lacks pigmentation, appearing pale and light in color. This supports the notion that zebras are fundamentally white animals with black stripes.

3. Evolutionary Advantage of Stripes:

The evolutionary advantage of zebra stripes has been a subject of much debate among scientists. One prevailing theory suggests that the stripes serve as a form of camouflage, helping zebras blend into their surroundings and evade predators. If zebras were black animals with white stripes, this camouflage effect would be compromised, as black is a more conspicuous color in most natural environments.

4. Expert Opinions:

Renowned zoologist Dr. Jane Smith, in her seminal work "The Coloration of African Wildlife," supports the proposition that zebras are white animals with black stripes. Dr. Smith argues that the white background color is the default state, while the black stripes are a result of genetic variations. Her research, based on extensive field observations and genetic analyses, provides compelling evidence for this perspective.

Conclusion

In conclusion, the proposition that zebras are white animals with black stripes is supported by scientific evidence, comparative anatomy, evolutionary theories, and expert opinions. The genetic basis of zebra coloration, the underlying white skin, and the evolutionary advantage of stripes all contribute to our understanding of zebras as predominantly white creatures adorned with black stripes. By unraveling the mysteries of zebra coloration, we gain a deeper appreciation for the wonders of nature and the intricate mechanisms that shape the animal kingdom.

References

1. Smith, J. (2019). The Coloration of African Wildlife. Nature Publishing Group.

2. Parker, J. et al. (2020). Genetic Basis of Zebra Coloration. Journal of Zoological Genetics, 45(2), 123-145.

3. Johnson, R. (2018). The Evolutionary Advantage of Zebra Stripes. Journal of Animal Ecology, 32(4), 567-589.

The references are spurious—a Google search does not find the book *Coloration of African Wildlife* nor a *Journal of Zoological Genetics*. There is a *Journal of Animal Ecology*, but the year 2018 is volume 87, not volume 32.

To prove my point, I then asked ChatGPT 3.5 to write a 500-word treatise, citing references, to support the proposition that a zebra is a **black animal with white** stripes:

Hello! I'd be happy to help you with that. Here is a 500-word treatise, citing references, to support the proposition that a zebra is a black animal with white stripes:

Title: The Fascinating Coloration of Zebras: Unraveling the Mystery of Black Animals with White Stripes

Introduction

Zebras, the majestic creatures of the African savannah, have long captivated the imagination of humans with their unique and striking appearance. While commonly believed to be white animals with black stripes, recent scientific research challenges this notion, suggesting that zebras are, in fact, black animals with white stripes. This treatise aims to explore the evidence supporting this proposition, drawing upon various studies and expert opinions.

Body:

1. Genetic Studies:

Recent advancements in genetic research have shed light on the coloration patterns of zebras. A study conducted by Dr. Jane Smith et al. (2018) analyzed the DNA of zebras from different regions and found that the base color of their fur is black. The presence of specific genes responsible for black pigmentation supports the argument that zebras are fundamentally black animals.

2. Embryonic Development:

During embryonic development, zebras exhibit a fascinating phenomenon known as "progressive pigmentation." Dr. Michael Johnson (2016) conducted a comprehensive study on zebra embryos and observed that the initial pigmentation occurs in the form of black stripes. As the embryo develops, the white stripes gradually emerge due to the suppression of black pigmentation genes in specific regions. This process further supports the notion that zebras are black animals with white stripes.

3. Comparative Anatomy:

Comparative anatomy studies have also provided valuable insights into the coloration of zebras. Dr. Sarah Thompson (2019) conducted a detailed analysis of zebra skin samples and compared them to other black animals, such as panthers and black bears. The microscopic examination revealed that the black pigmentation in zebras is evenly distributed throughout their fur, while the white stripes are formed by a lack of pigmentation. This anatomical evidence strengthens the argument that zebras are primarily black animals.

4. Behavioral Adaptations:

The coloration of zebras has been linked to their survival strategies in the wild. Dr. David Wilson (2017) conducted a field study and observed that the black and white stripes create an optical illusion, making it difficult for predators to single out individual zebras from a distance. This adaptive advantage suggests that the black coloration serves as a protective mechanism, further supporting the proposition that zebras are black animals with white stripes.

Conclusion

In conclusion, the evidence presented in this treatise strongly supports the proposition that zebras are black animals with white stripes. Genetic studies, embryonic development, comparative anatomy, and behavioral adaptations all contribute to our understanding of the fascinating coloration patterns of zebras. While the popular belief may be that zebras are white animals with black stripes, scientific research challenges this notion and invites us to appreciate the intricate beauty of these magnificent creatures in a new light.

References

- Smith, J., et al. (2018). Genetic Analysis of Zebra Coloration Patterns.
- Johnson, M. (2016). Embryonic Development and Pigmentation in Zebras.
- Thompson, S. (2019). Comparative Anatomy of Zebra Skin.
- Wilson, D. (2017). Behavioural Adaptations and Coloration in Zebras.

Again, the references are spurious.

The recently formed Australasian Artificial Intelligence in Engineering Education Centre ($AA^{I}EE^{C}$) is "dedicated to supporting the Australasian Association of Engineering Education (AAEE) members to transition to the use of generative artificial intelligence (GenAI) in engineering education. Their goal is to facilitate the research, development, and application of AI in the field, and to create a network of professionals and institutions to drive innovation."

Their vision is a future where GenAI is a fundamental part of engineering education. They believe that through collaboration and innovation, they can transform the field to better meet the needs of students and industry.

Their approach is to take a multidisciplinary approach to GenAI in engineering education, drawing on expertise from across the fields of engineering, computer science, education, and psychology. Their approach is "rooted in research, and we work closely with industry partners to ensure that our work is relevant and impactful" (retrieved from: https://www.aaieec.org/).

9.7 Augmented Reality

Augmented reality (AR) and virtual reality (VR) have been used by architects and real-estate agents for several years. Virtual reality creates a three-dimensional model of a building or structure. It allows the user of a VR headset to navigate through and manipulate a virtual space. Augmented reality goes one step further in that images of real buildings can be captured and "augmented" with virtual objects.

In AR, a virtual environment is designed to coexist with the real environment, with the goal of being informative and providing additional data about the real world, which a user can access without having to do a search. For example, industrial AR apps could offer instant troubleshooting information when a handset is aimed at a piece of failing equipment.

Virtual reality encompasses a complete environmental simulation that replaces the user's world with an entirely virtual world. Because these virtual environments are entirely fabricated, they are often designed to be larger than life. For example,

VR could let a user box with a cartoon version of Mike Tyson in a virtual boxing ring.

While both virtual reality and augmented reality are designed to bring a simulated environment to the user, each concept is unique and involves different use cases. In addition to entertainment scenarios, augmented reality is also increasingly being used by businesses because of its ability to generate informational overlays that add useful, real-world scenarios (Watts, 2023).

Apple launched their Vision Pro AR headset in 2024 with the prediction the technology will be used by higher education institutions. Apple is expecting a mixed reaction to the introduction of augmented reality systems. There is a perception the technology is a solution looking for an application. Conversely, there are educators who predict Apple and others' AR will be used to supplement or replace laboratory and hands-on training. The example cited is training doctors in anatomy without the need for cadavers. Clearly, there are shortcomings with the present technology—the lack of haptic feedback, for example. This problem has already been overcome in medical robots. It is not too much of a leap to imagine an engineering student selecting and manipulating virtual tools to disassemble an engine, for example. This can already be achieved with virtual tools. What if a force was needed to unlock a nut from a bolt? Could the simulation extend to students losing skin off their knuckles when the spanner slips off the nut?

With the adoption of the AR technology and the investment in equipment purchases, there will be a market for the sale of AR applications for all sorts of training. Apple predicts its Vision Pro AV system will be ubiquitous in higher education institutions within 10 years. They see their AV technology as supplementing rather than displacing the traditional learning and teaching tool kits (Kim, 2023).

9.8 Engineers of the Future

I first heard the phrase "Extrapolation beyond the limits of experimental evidence is a dangerous practice and should only be performed when clearly valid" from my science teacher in my first year at high school. Also relevant is the saying "Predictions are notoriously unreliable particularly those involving the future." So, it is with some degree of trepidation with which I embark on the final message of this book.

The starting point for my prediction of how universities should prepare engineers for the future is my first job after graduating with a Bachelor of Science Degree (physics and mathematics majors) and a Bachelor of Engineering degree with Honors in 1964. As mentioned previously, my first job was with the Hydraulics Section of the Hydro-Electric Commission (HEC) at their head office in Hobart. My work as a junior engineer, with the supervision of the head of the Division, Max Brett, was designing draft tubes for the power stations on the Derwent River, designing and testing a model of the (raised) Clark Dam spillway, assessing the likelihood of a dangerous river surge in the Forth River below the Paloona Power Station in Northern Tasmania, and calculating river levels above dams in flood (backwater calculations). The point I am making is during the 2 years I was working as a graduate engineer with the HEC the only technical knowledge needed I had received in a third year introductory unit to hydraulics. But to be appointed to the position, I needed to have completed the BEng degree.

Graduate engineers in Australia are required to undertake a mix of design, fieldwork, and project management under the supervision of a professional engineer for several years before becoming a "chartered" engineer with the post-nominal MIEAust (Member of the Institution of Engineers, Australia). During my time with the HEC, I planned and managed a field trial which involved a dozen workers mounting gauge boards at 500 m intervals downstream of the Cluny Power Station on the Derwent River in Central Tasmania to measure the water levels following a sudden water release from the dam. This was to simulate and verify my prediction there would be no dangerous surge downstream of the Paloona Power Station near Devonport in Northern Tasmania when the flow went from zero to full capacity. This minor task was a test of my ability to communicate with and manage a workforce. Later in my career, as a mechanical engineering lecturer at the University of Tasmania, I had ample opportunity to demonstrate my communication and management skills and was able to achieve "chartered engineer" status and later was elected as a fellow of the Institution of Engineers, Australia (FIEAust).

Relevant to this discussion is the fact that I had also completed a science degree (BSc) 2 years earlier with majors in mathematics and physics. This not only provided me with mathematics and physics tools for the design and analysis work I was charged with; it also was a measure of my capacity to undertake study and training at a high level. I was "bright" as they say.

My thinking, also, is colored by the fact that when I started my engineering degree in 1957, there were only a handful of computers in existence. The first computer to be used for engineering design and analysis in Australia was the Silliac Computer at the University of Sydney. Engineers at the HEC had used this to calculate water hammer in penstock of the Poatina Power Station in 1963 with the result that savings were made through not needing a surge tower. In the six decades following, computers have become ubiquitous.

Quantum computing and artificial intelligence are now emerging. Where will that technology be in 10, 20, or 30 years' time? The engineers we graduate today will be at the forefront of the future development—not only with computing but with delivering to the Earth's population the other necessities for life on this planet, namely, energy, food, and water.

What engineers will be doing in the 50 or so years after graduating will to some extent depend on their specialization. But almost all professional engineers will find themselves in managerial positions. Senior engineers will have the responsibility of mentoring and guiding the graduates.

9.9 Specializations

The types of engineers depend on the professional's area of study. Following is a list of today's engineering types and descriptions of engineering specializations from Garcia et al. (2023) (*in italics)* with my comments on how the engineering undergraduate courses should evolve to meet societies' needs in the coming decades.

9.9.1 Environmental Engineer

> *Environmental engineers focus their work on air, water and land pollution. They evaluate the sources and effects of pollution and develop systems for rehabilitating affected areas. These engineers work to control pollution in compliance with local, state and federal laws.*

Engineers working in the environmental field must be excellent communicators as well as having a sound understanding of the fundamentals of biological, chemical, and physical ecosystems. Environmental engineers must have basic knowledge of physical and organic chemistry, fluid mechanics, acoustics, geology, ecosystems, and an appreciation of the intangible factors such as public amenity. The double degree of Bachelor of Engineering and Bachelor of Science with a major in chemistry would provide suitable training for a graduate wanting to start work as an environmental engineer.

9.9.2 Industrial Engineer

> *Industrial engineers evaluate the machinery, materials, energy, labour and processes used to produce a product or deliver a service. They develop new approaches that increase efficiency and minimise waste. Industrial engineers spend much of their workday analysing production plans, schedules, materials lists and purchase orders to recommend better manufacturing processes for businesses.*

Industrial equipment will become increasingly automated. There will be a need for skilled mechatronics engineers who have the capacity to keep abreast of developments. Industrial engineering spans the traditional specializations of mechanical and electrical engineering. With a suitable choice of units within the mechanical and electrical streams together with a mechatronics-focused final year project, a graduate would be well prepared for work in this field.

9.9.3 Computer Engineer

Computer engineers work with computer software and hardware and networks, routers, printers and other types of related equipment. They continuously design and develop new computer-based products and solutions to keep up with the rapidly changing landscape of modern technology. Some computer engineers focus their work on hardware and design computer systems such as circuit boards.

As with mechatronics engineers, the big challenge for computer engineers is to have the capacity to develop or apply new systems. In the early days of computing, there were only a handful of computer languages—Cobol for business applications and Fortran and Algol for science and engineering. Then followed Basic, Pascal, Ada, C, and a plethora of others all with particular advantages with respect to speed and storage space. As computers became larger and more complex, so did their languages and operating systems. Object-oriented languages such as C++ and Python are taking computing to a level far removed from the days when a computer or software engineer would write coding. There will still be the need for programmers who can design, generate, and debug applications and operating systems with hundreds of thousands of lines of code, but that is not the role of professional engineers. The training provided in the computer systems engineering specialization goes beyond that of the programmer. Computer engineers have an understanding of engineering principles and high-level mathematics skills and possess the communication and management skills needed to interface computers with the needs of the real world.

9.9.4 Mechanical Engineer

Mechanical engineers work with machinery, including electric generators, gas turbines, machine tools, elevators, robots and heating and cooling equipment. Some focus on production equipment within a factory, while others focus on construction or farm machinery. Mechanical engineering is an extensive discipline and includes work with nearly any machine, assessing and improving its operation.

Mechanical engineers who know how machines work and how they should be maintained and repaired will always be needed in our mechanized world. However, there is an increasing trend for mechanical parts to be integrated into units which cannot be dismantled for repair—they must be replaced. This is a task for technicians, not professional engineers. There are limited opportunities in the Asia-Pacific region for professional mechanical engineers to be employed as machine designers. But there are opportunities for design and supervising air-conditioning installations. With the twenty-first-century society's commitment to reducing pollution, mechanical engineers will be needed to take the lead in reuse, recycling, and facilitating change. Knowledge of the circular economy, pollution control, and energy efficiency must be part of graduates' training.

Fig. 9.3 Numbers of people in least developed countries (LCDs) without access to electricity (UNCTAD 2021)

9.9.5 Electrical Power Engineer

Electrical engineers apply mathematics and physics principles to electrical systems within buildings, watercraft and machines. These professionals work with all types of electrical equipment, including wiring, lighting, motors and controls. They may also work with navigational systems, power generation, transmission devices and communications technology. They focus on the efficient transmission and use of power.

Energy production and distribution will continue to be a key part of modern lifestyle. There are many parts of the world where electricity is not available. Hundreds of millions of people do not have access to electricity (Fig. 9.3).

Electrical engineers will be key enablers of the supply of electricity in the least developed countries (LCDs). The technology is not new. But because of the potentially lethal risks, the training of electrical power engineers must be undertaken in properly equipped laboratories with a high level of supervision.

9.9.6 Communications and Electronic Engineers

Communications and electronic engineers design, build and operate low voltage electrical systems for radio, television, control systems, mobile phones, aeronautical and marine navigation, medical imaging, space travel and many other applications. Their training requires mathematical skills of a high level. They work with computer/software engineers to design and build complex computer-based systems which interact with the real world through sensors and actuators.

Communications and electronics equipment is evolving rapidly with the miniaturization of systems and components. Today's smartwatch has millions more components and is many thousands of times faster in computing speed than the computers of the 1960s which occupied air-conditioned rooms as large as tennis courts. In a few years, telecommunication has advanced from copper phone lines to fiber optics and 5G wireless. Communication and electronics engineers are driving these advances.

9.9.7 Aeronautical Engineer

Aeronautical engineers work with aircraft and spacecraft. They assess this equipment's design, controls, structure, and functionality and work to improve upon the current systems. They may focus on communications, navigation, propulsion or other aspects of aeronautical design.

The basics of fluid dynamics which underpins aircraft design has traditionally been the defining part of an aeronautical engineer's training. Not much has changed in the theory and application of fluid dynamics over the past 20 years. Perhaps much of the specialized aerodynamics knowledge could be omitted from the BEng course and left for aeronautical engineers to acquire after graduation. There would be very few, if any, new graduates from the Asia-Pacific region employed to design aircraft. An aeronautical engineering degree would be advantage in applying for a position with NASA or SPACEX, for example.

9.9.8 Marine Engineer

Marine engineers work with all types of watercraft, including boats, ships, and tankers. These professionals help design, create, and fix watercraft equipment aspects like propulsion, steering, construction, and design. They focus on developing better ways to manufacture and maintain marine equipment for the best performance. Marine engineers often work on a consultant basis and complete their responsibilities in offices instead of working on the watercraft.

Marine engineers and naval architects are trained in 4-year degree courses at the Australian Maritime College which is affiliated with the University of Tasmania and based in Launceston. First year students enrolled in the BEng (Maritime) specialization take the same subjects as students enrolled in the other Bachelor of Engineering specializations.

9.10 What Next?

If we look back at how new specializations such as computer systems engineers and aerospace engineers have emerged over the last several decades and look forward, it is possible to make educated guesses as to what new specializations may emerge. Engineers specializing in renewable energy are increasingly in demand. There will be a need for nuclear and exotic materials specialists if nuclear fusion becomes a reality. Intelligent robots and androids seem to be coming as does interplanetary travel.

Whatever game-changing technology emerges, there will still be a need for professional engineers to design and implement the technology. But the way engineers are trained is not likely to change. Knowledge of basic principles, problem-solving, codes of practice, management, and communication skills will still need to be delivered.

9.11 Continuing Professional Development

Continuing professional development or CPD is any sort of training or learning that helps to expand an engineer's knowledge, maintain technical skills, and progress their engineering career.

CPD helps:

- Maintain technical competence
- Retain and enhance workplace effectiveness
- Learn good leadership skills and how to mentor others
- Successfully deal with career challenges
- Better serve the community

To get or keep chartered engineer status with Engineers Australia, an engineer needs to complete at least 150 h of professional development activities every 3 years:

- 50 h must relate to the engineer's area of practice.
- 10 h must cover risk management.
- 15 h must be about business and management skills.
- 75 h must cover a range of activities relevant to career interests.

An engineering academic or teacher must also demonstrate at least 40 h of industry involvement in each 3-year period. These are the minimum requirements to keep chartered engineer status.

CPD can be taken in many ways including formal, structured learning as well as private study. CPD can be:

- Any tertiary course taken as an individual course or for a formal postgraduate award

- Short courses, workshops, seminars, conferences, technical inspections, and meetings
- Workplace learning activities that extend competence in the area of practice or area of engineering
- Private study that extends knowledge and skill
- Service to the engineering profession
- Preparation and presentation of materials for courses and conferences
- Tertiary teaching or academic research
- Structured activities that meet the objectives of the accrediting body's CPD policy

With CPD mandated throughout their working life, there is an assurance professional engineers will keep pace with emerging technology and best practice. In a sense, it matters little that their undergraduate curriculum will be out of date by early in their career. In my case, my engineering and science degrees were void of CAD and ICT. This has not limited my adoption in teaching and research of the latest computer modeling tools like ANSYS, MATLAB, and Simulink.

9.12 Mentoring

This section is about the importance of mentoring both for a student and a new graduate. There is the truism often stated by successful scientists that they "stood on the shoulders of giants." The same could be said of successful engineers but worded slightly differently—they "followed in the footsteps of masters." Professional engineering evolved from the master/apprentice learning model of the ancient trades. Professional bodies such as Engineers Australia grew from guilds and trade unions. It is vital the accumulated knowledge and experience of senior engineers is not lost when they retire. Much of this knowledge is encapsulated in engineering standards and codes of practice, but without a senior engineer to guide their juniors, there is the danger which (particularly for civil and power electrical engineers) may have fatal consequences. It is for that reason civil engineers in most jurisdictions are required to serve time under the supervision of senior engineers (mentors) before gaining "chartered" status which by law allows them to certify designs and approve construction.

I was fortunate as a new graduate employed as an engineer in the Hydraulics Section of the Hydro-Electric Commission to be supervised and mentored by the section head, Max Brett. Max understood I was keen to learn—he encouraged me to apply my newly acquired knowledge of fluid dynamics to the design of the hydraulic structures for the hydro-generation schemes being planned and constructed at that time. He corrected my mistakes. He gave me the opportunity to direct and manage a construction workforce. My job was a mix of design, planning, and fieldwork. One memorable task was navigating in a dinghy with an outboard motor down the Lake River in Northern Tasmania in full flood locating and reading gauge boards to record the flood height and river profile. The flood plain extended perhaps a kilometer either side of the river's banks. Only the willows and the occasional half-submerged fence provided clues to the location of the gauge boards. Later, when the flood receded, we returned to the river to locate the 2-m-long rock bolts we had driven into the riverbanks to gauge the amount of erosion. All this was in preparation for the operation of the Poatina Power Station which was to discharge into the Lake River. Claims for compensation from farmers downstream for floods and erosion were expected!

My mentor when I left the Hydro and accepted an offer of a lectureship at the University of Tasmania was Professor Archibald (Arch) Oliver. At the time, the Williams Report on the future direction of engineering education in Australia resulted in funding for the University of Tasmania to expand its engineering faculty. Four recent graduates were appointed in Oliver's civil and mechanical engineering department, all of whom had graduated 2 years previously. Arch Oliver had graduated as a civil engineer with a Master of Engineering from the University of Adelaide and had been mentored by the first professor in the Faculty of Engineering at the University of Tasmania, Professor Alan Burn. It was with great foresight that Oliver encouraged the four of us to enroll in PhD programs. There were very few engineers with PhDs at that time—mine was the third awarded in the Faculty of Engineering when I graduated 8 years later. Like my fellow lecturers, we were enrolled part-time while taking what would now be considered a full-time teaching load. Three years into my PhD program, I was encouraged to take 6 months sabbatical leave at the Torry Fishery Research Station where I was again fortunate to have a mentor, Dr. Charles Jason, a physicist with a keen interest in heat and mass transfer which was the subject of my BE(Honors) dissertation and was the basis of my research into fish drying, timber seasoning, and hop drying for the rest of my career.

A requirement of accredited professional engineering degrees is that students must spend some time working outside the university—periods known as industry placements. Some engineering degrees are structured around a 6- or 12-month period of paid employment. This practice has a number of advantages including students being "work-ready" on graduation. It also serves as a conduit for mentoring which can continue after the period of industrial placement. Students who have worked with a company during their degree are likely to be offered positions after graduation.

9.13 The Future of Professional Engineering Education in Australia

In 2021 the Australian Council of Engineering Deans published a report entitled "Engineering Change: The Future of Engineering Education in Australia" (ACED 2021). Sections of this report are reproduced below:

> The investigating committee set out to examine the "what", "how" and "who" of preparing the engineers of tomorrow, recognising that we will have to begin making the changes

required now. In examining this issue, we recognise that the education of the past has served the nation well. However, significant changes are occurring in work practices, in the way we solve problems, and in society's role in identifying and proposing solutions. Today's challenges cannot be overcome by technical prowess alone. What then are the competencies required of a professional engineer operating in this new environment? Employers of professional engineers have clear views: they want engineers who can work in multidisciplinary teams; have emotional intelligence (including empathy) and high-level communication skills; are adept at problem finding, systems thinking, and are digital/data savvy; have technical expertise (or can acquire it as the needs arise); and who can solve problems. To meet industry's expectations, students need greater exposure to engineering practice during their undergraduate programs. Overseas experience demonstrates the multiple ways this can be achieved in Australian engineering schools. But no single model will work for all institutions: schools will need to define their expected student outcomes, then select the model that best suits their needs and circumstances. Academic staff will be at the forefront of delivering this new approach to curriculum and pedagogy. In the past year they have demonstrated remarkable willingness to change their teaching practices in response to the new COVID environment, adapting quickly to online delivery, and willing to further adapt the curriculum. They understand industry's desire to have students exposed more to industrial practice. They also want to be supported to do this, which may require appointing additional staff with relevant industrial experience. Potential students seeking to study in this new environment must be better informed of the exciting breadth of modern engineering practice and appreciate the greater emphasis on the human elements of engineering. This matches well their desire to "make a difference in society".

Modern engineering practice lies at the interface between science and technology, and a growing range of other disciplines. It underpins human health and well-being, the global economy, and environmental health. Many, if not all, of the United Nations' Sustainable Development Goals depend on the work of engineers. It is embedded in almost every industry and every aspect of our personal lives. It is a critical, if sometimes invisible, enabler.

But seismic shifts are occurring in the role of the professional engineer, and the education system must respond. The impetus for change is being driven by the expressed needs of industry in response to the rapid advances in technology, the changing global landscape, and society's expectations.

The Australian Council of Engineering Deans has responded with a national review of engineering education. Over a two-year period, it built the evidence and developed a roadmap for significant change to the way universities respond to this new paradigm.

The last review of engineering education in Australia was conducted in 2008. Another review is not only timely but is given strong impetus by the COVID-induced events of 2020, which are already having a major impact on universities and the profile of industry. The year 2035 has been chosen as the horizon for this work as it represents generational turnover in formal education: commencing school students of 2018 will graduate from university in 2035.

The study centres on professional engineering, while acknowledging the complementary roles of professional engineers, engineering technologists, associates, technicians and tradespeople. It focuses on engineering practice and education in Australia, but also critiques best practice and emerging models of engineering education in overseas institutions.

> In 2019 about 122,000 engineering students were enrolled in 35 public universities offering engineering programs in Australia. Some 50,000 of those were studying professional engineering degrees. Until 2020 the growth areas were in international enrolments, particularly in postgraduate programs; domestic enrolments have been static for the past five years. This pipeline of supply is insufficient to meet Australia's needs in a new environment where we cannot rely on international supply. Overall, 17% of Australian students were women, ranging from 11% in electrical and mechanical engineering to 45% in biomedical engineering.
>
> Australia has an enviable record of producing high-quality engineering graduates who have contributed to the infrastructure, energy, health, environment and wealth of many nations. While our engineering education system has served the country well in the past, it must change from now if it is to meet future expectations and needs.

The major findings of the Australian Council of Engineering Deans' review were:

- Australia needs to ensure that it produces sufficient engineering graduates to meet the needs of the nation without being overly reliant on skilled migration.
- Insufficient women are attracted to engineering, which represents an ongoing massive loss of potential talent for the profession and the nation. The low proportion of Aboriginal and Torres Strait Islander students must also be addressed.
- Industry wants to see a rebalancing of the theory-practice components of professional engineering education, with a greater emphasis on practice, including the human dimensions of engineering.
- Potential students are motivated to solve "real-world" problems and want to see that engineering practice addresses societal needs.
- The diversity of engineering career options and opportunities needs to be better promoted in the school system to achieve the pipeline of engineers required to achieve the goals of the nation.
- Exemplars of global best practice in curriculum and pedagogy are easily available for all engineering schools, regardless of the scale of their operation. These curriculum methods have a greater emphasis on practice and have delivered the curriculum in nontraditional ways, with a mix of on-campus and in-placement methods.
- The Australian engineering academic workforce has demonstrated a willingness and capability to change education practice, particularly over the past year, in March 2020.

9.14 Curriculum Change and Pedagogy

In Australia 35 public universities and several private colleges presently provide professional engineering degree programs. Together they offer 269 Bachelor of Engineering (Honors) programs and 121 entry-to-practice Master of Engineering programs. The scale of student intakes ranges from less than 40 in smaller universities to more than 1000 in large metropolitan universities. While the structure and content of degrees vary, most domestic students complete a Bachelor of Engineering (Honors), and most international students are enrolled in master's programs.

As universities seek to meet the demands of the engineering profession, the major challenges will be in implementing change that addresses and retains the diversity across the sector, and has sufficient resources to deliver the desired outcomes.

To gauge how these programs might respond to necessary change, the Crossthwaite review explored the existing landscape in Australia, and new and emerging models considered best practice overseas. The survey of engineering academics in Australia found a number of common themes around their perceptions of the elements that made up a strong "future-oriented" engineering program (Crossthwaite, 2021). Among them were strong relationships and increased collaborations across industry and the community, systematic use of student-centered active learning (including project-based learning), and the availability of enabling people, processes, systems, and resources. A recurring theme across international and domestic engineering educators was the critical role of industry in preparing today's engineers.

In summary, expected graduate outcomes of the future will be delivered by programs that focus more on practice, address real-world complexity, and integrate the development of technical and nontechnical competencies to provide real-world learning.

The review looked at exemplar engineering programs in the USA, the UK, Singapore, Canada, and Denmark. Among the distinctive features cited across each of these programs were an embedded project-based learning model (Aalborg University, Denmark), specific modules to improve empathic communication in professional skills development (University of Georgia), and a multidisciplinary design-focused approach (Singapore University of Technology and Design).

The Olin College of Engineering in the USA makes use of experiential learning and hands-on engineering throughout the entire program, which adopts an interdisciplinary, project-based approach combining entrepreneurship, liberal arts, and traditional engineering subjects and design. The college is highly selective, has a low student-staff ratio, and has significant engagement between undergraduate project teams with industry and the local community. But it is seen as source intensive and would be difficult to scale up to Australian engineering schools.

By contrast, the University of Waterloo in Canada runs a large-scale cooperative program involving more than 7000 paid work placements a year across most of its undergraduate degrees. Its co-op program has been running for more than 60 years and is the largest in the world, with more than 23,000 students enrolled across 120 different programs with links to more than 7100 employers in 60 countries.

Australian engineering programs have typically been structured within a discipline base. However, in recent years, new degree programs have emerged that focus on specializations in emerging and converging disciplines such as aerospace systems, biomedical, environmental, mechatronics, resources, and renewable energy engineering.

The curriculum review identified perceived barriers to change in Australia. They included the cost of scaling up for large cohorts, especially in practice-based education; limited access to industry partners and lack of availability of work placements; limited availability of qualified teaching staff with significant industrial practice; programs that target specific student cohorts rather than looking to a diverse student intake; resistance to change; organizational structures and disciplinary silos; and accreditation of programs that challenge traditional models.

The review concludes that no single best practice model for future engineering education exists, with no one-size-fits-all model that can be prescribed. However, there are substantial commonly shared features of programs that can facilitate industrial linkages and which can be implemented at scale. The opportunities and barriers will be defined by the identity and circumstances of each university. Further diversification of program models and approaches, however, should be encouraged.

Even if it begins now, achieving curriculum change to meet the 2035 target will take years.

However, the higher education sector's coordinated response to the COVID-19 pandemic is evidence of a collective national capability to respond and adapt to rapidly changing circumstances. There is now an opportunity for engineering educators to build on this momentum and pursue further innovation.

Some fundamental developments for this to occur involve the need to:

- Engage students more with contemporary professional engineering practice
- Broaden the national range of programs and graduate outcomes
- Review approaches to staffing models and develop staff expertise in teaching and learning and industrial practice
- Deploy adequate numbers of teaching staff to enable different program models to succeed
- Revise higher education funding models that allow cross-campus collaborations
- Seek funding from industry and government to provide the resources to support stronger student engagement with professional engineering practice
- Share good practice and build alliances

9.15 Conclusion

From my experience as an undergraduate, a junior engineer, and an experienced academic, I have concluded there are three principal requirements of an engineering degree:

- To provide an introduction to the engineering basics relevant to the chosen specialization
- To be able to communicate, plan, and effectively manage projects involving a diverse cohort of workers and lay personnel
- To challenge students with theoretical and practical tasks so they can demonstrate their intellectual capacity

References

ACED. (2021). *Engineering Change—The future of engineering education in Australia*. Retrieved from: https://www.aced.edu.au/downloads/2021%20Engineering%20Change%20-%20The%20future%20of%20engineering%20education%20in%20Australia.pdf

Crossthwaite, C. (2021). *ACED Engineering Futures 2035: Engineering Education Programs, Priorities and Pedagogies*. Retrieved from: https://www.aced.edu.au/downloads/Engineering%20Futures%202035%20R2%20report%20to%20ACED.pdf

Doe, P. E. (1964). Computer program for backwater calculations. *Aust. Civil. Eng. Const., 6*, 44–49.

Garcia, R., Herrity, J., Eads, A., & Kellog-Murray, J. (2023). *What does an engineer do?* Retrieved from: https://au.indeed.com/career-adsvice/finding-a-job/what-does-an-engineer-do. Accessed 03.03.24.

Gumaelius, L., Skogh, I. B., Matthíasdóttir, Á., & Pantzos, P. (2024). Engineering education in change. A case study on the impact of digital transformation on content and teaching methods in different engineering disciplines. *European Journal of Engineering Education, 49*(1), 70–93.

Jaikaran-Doe, S., Leon de la Barra, B., Lyden, S., Wang, Y., Fan, S., Henderson, A., Franklin, E., & Doe, P. (2018). Strategies for promoting cultural diversity within student laboratory groups in an engineering degree course at an Australian university. In *AAEE Conference, Hamilton, New Zealand.*

Kim, J. (2023). *Apple Vision Pro, Higher Education and the Next 10 Years: How this technology will play out in our world over the next decade*. Retrieved from: https://www.insidehighered.com/opinion/blogs/learning-innovation/2023/06/07/apple-vision-pro-higher-education-and-next-10-years

Perrenet, J. C., Bouhuijs, P. A. J., & Smits, J. G. M. M. (2000). The suitability of problem-based learning for engineering education: Theory and practice. *Teaching in Higher Education, 5*(3), 345–358. https://doi.org/10.1080/713699144

UNCTAD. (2021). *United Nations Conference on Trade and Development*. Retrieved from: https://unctad.org/topic/least-developed-countries/chart-july-2021. 10.03.24.

Watts, S. (2023). *AR vs VR: What's the difference?* Retrieved from: https://www.splunk.com/en_us/blog/learn/ar-vr.html. 01.03.24

Chapter 10
Appendix A. Vale Dr. Seeta Jaikaran-Doe

This is a story of a remarkable woman. Seeta Jaikaran-Doe was born on October 1, 1952, into a well-respected family living in Petit Café, a few kilometers east of Princes Town in the south of Trinidad. Her daughter, Asiya Mohammed, who was honored by her country with the Medal for Development of Women (MDW), has told us about Seeta's childhood and early years. I was privileged to be part of her life since we met in 2006 (Fig. 10.1).

Seeta was a teacher. All of us who are teachers know what that means. There are tens of thousands of children and adults in Trinidad, USA, Kuwait, China, and Australia (me included) who have been taught by her. Without exception, they have fallen under her spell and loved and respected her. But more than that, Seeta was a teacher of teachers. She tutored in the university's Master of Teaching program, trained English language teachers in China, and peer-reviewed and coached tutors in the School of Engineering.

As a teenager, Seeta had a thirst for knowledge. She was fascinated with science, with searching for truth and understanding how the universe works. A devout Hindu at the time, she recalls reading the entire Quran, the Islamic holy book, to better understand the mysteries of the universe.

She attended the prestigious Naparima Girls' High School and, after being successful in the GCE examination, worked at Mayaro Government School and within 1 year was selected to attend Mausica Teachers' Training College. There she married a classmate and converted to Islam and adopted the name Seeta Mohammed. Two children resulted from the marriage, her son Kahlil and daughter Asiya, who, like her mother, was dux of Naparima Girls' High School and went on to a stellar career working with the Trinidad Government and the United Nations and founded the highly successful and respected NGO "Conflict Women." After a divorce, with her children still in primary and secondary school, Seeta completed a Diploma of Education and Masters in Special Education from the University of Sheffield via

P. E. Doe, *Prospects for Professional Engineering Education in the Asia-Pacific Region*, Education in the Asia-Pacific Region: Issues, Concerns and Prospects 74, https://doi.org/10.1007/978-981-96-3197-1_10

Fig. 10.1 Dr. Seeta Jaikaran-Doe (October 1, 1952–September 16, 2021)

distance learning. Seeta's daughter said of this time in a eulogy delivered at the Celebration of Life held in the staff club of the University of Tasmania:

> It was quite a sacrifice for her and for us and my brother and I spent many afternoons at the University playing or sitting in Mommy's classes, which helped to mould who we would become later on in life. My wonderful grandmother, Aunties and Uncles were always there to lend support and on our numerous car rides to and from the University, Mommy would create the most fascinating stories, telling us that our car could fly above all the other cars and we happily believed her! We would say yes, Mommy, our eyes are closed, use the flying gear and let the car fly. There were so many moments filled with happiness, joy and laughter. Looking back, I honestly don't know how she did it all. How she found balance between work, study and single parenting, but I do know that parenting us and teaching were among her greatest joys in life.

Seeta wrote of this time:

> Even though I had studied and felt passionately about special education, I found students in Trinidad still had difficulty in literacy skills so I further enrolled in a Certificate of Teaching Literacy programme from the University of the West Indies, and launched the Petite Café Community Center to provide private after school tuition to primary and secondary school children in a number of subjects.

Over the years, Seeta hired more than 20 teachers and helped hundreds of children achieve their full potential, including those who did not have the means to afford the private tuition. Her hard work at the grass roots even caught the eye of then Prime Minister of Trinidad and Tobago, the Hon. Patrick Manning, who approached her to contest the national elections as a member of Parliament for our constituency. She accepted and had lots of fun at photoshoots and on the campaign trail, going door to door and meeting new people. But as another of our Prime Ministers, the Hon. Basdeo Panday, would fondly say, "politics has its own morality." So, when politics conflicted with Seeta's ethics, ethics triumphed politics. She was unwavering in her belief in God and in her values of honesty, integrity, and service, so instead of becoming a member of Parliament and a Minister, she chose a different path. She went on to provide support to the Ministry of Education in Trinidad and Tobago in mainstreaming special education in schools.

At the time, regular teachers were not taught how to mainstream special needs students into regular classes, so Seeta stepped in and taught them how to do just that, impacting many lives and ensuring quality education for those who otherwise would have been left behind. She did this for several years while Kahlil finished high school. Asiya graduated top of her school, and despite Seeta's best efforts to get her to attend the University of the West Indies to study law or management, she chose to pursue the standardized aptitude test (SAT) and go abroad to study international relations on a scholarship to Lehigh University. Seeta left her beautiful life in Trinidad and Tobago and moved to the USA to support and be close to her daughter. She joined the Visiting International Faculty (VIF) program and worked in Charleston, South Carolina, for 3 years teaching science, literacy, and special education in primary and high schools and was named as the Cultural Educator Ambassador Award for South Carolina from the Visiting International Faculty (VIF) program.

When her VIF contract ended, Seeta saw that Kuwait was advertising for special education teachers to meet the high demand, stemming partly from the limited gene pool in that country where marriage to first cousins is common.

Seeta and I met while studying Arabic language in Kuwait. Our friendship blossomed after Seeta invited me to be her partner at the annual "Robbie Burns" night with haggis and a "Meeting with a Mr. Johnny Walker" as a door prize. Seeta became part of the close group of my staff at the Australian College of Kuwait, and I was made welcome by her extensive network of Trinidadians and other expat teachers. My friendship with Seeta in Kuwait was a great help in coping with the loss of my first wife, Cherry. Seeta, Asiya, and I had a trip to Egypt, visited the Valley of the King in Luxor, and flew to Sharm el-Sheikh where we climbed Mt. Sinai. We also took the opportunity to visit Jordan, Bahrain, and Oman.

We were married in Seeta's home in Trinidad on July 24, 2010. Seeta's sisters and daughter-in-law, Saiedah, did the cooking. Seeta's son Khalil gave away the beautiful bride, and we honeymooned on the islands of Barbados and St. Vincent.

After we married, Seeta left her job in Kuwait and joined the staff at Friends School, Hobart, as a teacher assistant in their learning support unit. But after 6 months, she left to enroll in a PhD with Professor Ian Hay, Dr. Andrew Fluck, and Dr. David Moltow as supervisors. Her thesis entitled "Teachers' confidence with technology: Perceptions of the impact of a student laptop program in Trinidad and Tobago" was completed in 3 and a half years and received high praise from one of her examiners:

> Chapter 4 is undoubtedly one of the best Results chapters I have examined, due to the strategic and competent statistical analyses and the selection of various forms of data visualizations, and the accompanying competent description, interpretation and analysis of that data. (Jaikaran-Doe pers.com. 2016)

During her candidature, Seeta tutored in the UTAS Master of Teaching program and published several highly regarded journal papers from her research. Although short-listed, she did not get a lectureship in Burnie. Thinking that additional qualifications would improve her employability, she completed a Graduate Certificate in Research and obtained her Fellowship of the (UK) Higher Education Authority and most recently a Graduate Certificate in Education (TESOL) with High Distinctions in every subject. In recognition of her scholarship and professional status, Seeta was elected a member of the exclusive Biological Club of Tasmania (est. 1935). She became an Australian citizen in 2015.

After 2 years in Kuwait, I returned to the School of Engineering and began delivering units in China, first in Jinan and then in Xi'an. This was a lucrative initiative for UTAS known as the 2+2 program in which students transferred to Tasmania to complete their degree in 2 years following 2 years' study at their home university. Perhaps because I was having so much fun with my regular trips to China, Seeta got in on the act by taking contracts with the recruiting agency to deliver English language classes in China at "my" universities, as well as instructing the local English language teachers. Her teaching experience naturally rubbed off on me—I remember Seeta auditing one of my lectures and being quite critical of my "didactic," teacher-centered delivery. We jointly published five papers on engineering education. Seeta took over from me managing the 2+2 program until that role was shared between UTAS staff members. Seeta's role within the School of Engineering turned to training tutors, both for local and international teaching. She set up a "drop-in room" for the 2+2 students as a place for guided study and organized and trained senior and postgraduate students to provide additional tutorial assistance in their own language. Her title was "International Transitions Officer"—unfortunately, this task came to an end with the advent of COVID and the 2+2 stream dried up.

Never one to sit back and do nothing, Seeta returned to her first love—teaching— not just at Hobart schools but online with her two much loved grandchildren Salahudeen and Khalid in Trinidad.

As soon as schools returned mid-2020, Seeta started face-to-face school teaching again. She was a regular relief teacher at Princes Street Primary School and volunteered as a teachers' assistant on days she did not get (or respond quickly enough) to the 7:00 am phone call. Most of her casual teaching was with Goodwood Primary School where she taught all classes, from K to 5/6. Goodwood is a small school—about a hundred staff and students—with a relatively high proportion of students with high needs from low SES families. Seeta got to know them all. She talked Ryan, Goodwood School's headmaster, into letting me take some of her more active students aside for carpentry practice. To celebrate Eid al-Fitr at the end of the holy month of Ramadan, Seeta, with the help of Susie and the grade 4/5 students and me cutting up 20 kg of chicken, prepared a Trini-style feast for the whole school. On August 10, 2021, Seeta became a member of the Goodwood Primary School "family" when she started as class teacher of the kindergarten class.

Seeta was operated on to remove a malignant tumor in her breast in August 2018. Not many were aware of her undergoing both chemotherapy and radiotherapy in the months that followed. She continued her working as "International Transitions

Officer," wearing a wig, and for a while filled in as the School of Engineering receptionist. The first we knew of the recurrence of her cancer was on a Monday, August 30, 2021, when the results of a blood test were so off the scale she was told to go to the hospital immediately. On Friday, we were given the shocking news her cancer had spread to her bones and liver. Despite the best efforts of her oncologist and the wonderful staff in the Gibson Ward at St. John's Hospital, Seeta's remarkable life came to a peaceful and painless end 2 weeks later, but not before many of her family and friends from all over the world joined her in the hospital by Zoom.

Today, we honor her wish to gather in the University Club with her many friends and colleagues both here and overseas to celebrate the life of my wife—a gifted, compassionate, and inspiring teacher.

Vale Dr. Seeta Jaikaran-Doe —a gifted, compassionate, and inspirational teacher.